Assessing Revolutionary and Insurgent Strategies

THE SCIENCE OF RESISTANCE

I0092931

Summer D. Agan, Lead Author

Johns Hopkins University Applied Physics Laboratory (JHU/APL)

Contributing Authors:

Jonathon Cosgrove, JHU/APL

Robert Leonhard, JHU/APL

Guillermo Pinczuk, JHU/APL

Idean Salehyan, University of North Texas

United States Army Special Operations Command

The Science of Resistance is a work of the United States Government in accordance with Title 17, United States Code, sections 101 and 105.

Published by Conflict Research Group.

First published by USASOC in 2019

ISBN: 978-1-925907-31-5

CONFLICT
RESEARCH
GROUP

ASSESSING REVOLUTIONARY AND INSURGENT STRATEGIES

The Assessing Revolutionary and Insurgent Strategies (ARIS) series consists of a set of case studies and research conducted for the US Army Special Operations Command by the National Security Analysis Department of the Johns Hopkins University Applied Physics Laboratory.

The purpose of the ARIS series is to produce a collection of academically rigorous yet operationally relevant research materials to develop and illustrate a common understanding of insurgency and revolution. This research, intended to form a bedrock body of knowledge for members of the Special Forces, will allow users to distill vast amounts of material from a wide array of campaigns and extract relevant lessons, thereby enabling the development of future doctrine, professional education, and training.

From its inception, ARIS has been focused on exploring historical and current revolutions and insurgencies for the purpose of identifying emerging trends in operational designs and patterns. ARIS encompasses research and studies on the general characteristics of revolutionary movements and insurgencies and examines unique adaptations by specific organizations or groups to overcome various environmental and contextual challenges.

The ARIS series follows in the tradition of research conducted by the Special Operations Research Office (SORO) of American University in the 1950s and 1960s, by adding new research to that body of work and in several instances releasing updated editions of original SORO studies.

RECENT VOLUMES IN THE ARIS SERIES

ACKNOWLEDGMENTS

The editor would like to thank Dr. Idean Salehyan for his thoughtful review of this work. Any errors remain those of the editor. After conducting his review, Dr. Salehyan helpfully contributed the section on International Dimensions in chapter 2 to address gaps he identified in the initial draft.

TABLE OF CONTENTS

LIST OF ILLUSTRATIONS

Credits:

Figure 1. Continuum of categories of resistance. Erin Hahn, ed., *ARIS Legal Implications of the Status of Persons in Resistance* (Fort Bragg, NC: USASOC, n.d.)

Figure 2. ARIS proposed states for phasing construct analysis. W. Sam Lauber, Steven Babin, Katherine Burnett, Jonathon Cosgrove, Theodore Plettner, and Catherine Kane, *Understanding States of Resistance*, Draft (Fort Bragg, NC: USASOC, 2016).

Figure 3. Quadrant of key features of a state. Adapted from Charles Tilly, *The Politics of Collective Violence* (New York: Cambridge University Press, 2003), 48.

Figure 4. Equivalent response model. Guillermo Pinczuk, ed., *ARIS Threshold of Violence* (Fort Bragg, NC: USASOC, forthcoming).

CHAPTER 1.
INTRODUCTION: WHAT IS RESISTANCE?

The post-Cold War world has challenged the paradigm of Western great powers that have dominated world affairs since the Peace of Westphalia in 1648, which granted permanent and special status to sovereign states. Resistance movements targeting established governmental authority have existed since antiquity, but the prominence of internal conflict in world affairs has grown in the twenty-first century as civil wars have replaced interstate wars in frequency. Conventional warfare, though still relevant, demonstrated its limits in Vietnam, Iraq, and Afghanistan, and the prevalence of insurgencies, coups, popular uprisings, and revolutions clearly demonstrates that future threats are likely to include a complex brew of irregular conflict centered on resistance movements. Preparing to meet such a challenge requires a disciplined approach to understanding resistance movements.

In the latter stages of the Cold War and for the decade following it, the US military establishment rose to new levels of performance in joint warfare and set the standard for conventional combat power. From the messy conclusion of the Vietnam War through shock of the 9/11 attacks, the American military invested in the paradigm of integrated land, sea, and air warfare aimed at destroying enemy states' armed forces, their capacity to generate armed forces, and in some cases, the offending regimes themselves. The assumption that underlies American "traditional warfare" is analogous to a social pact in which the states involved, along with the populations over which they exercise sovereignty, will acquiesce in the military outcome and accept the subsequent political decisions. Indeed, this form of warfare attempts to keep noncombatants out of the fight and instead focuses on defeating enemy armed forces and seizing terrain.

Throughout this period, the joint forces were aware of another kind of conflict called "irregular warfare." US history includes many episodes of irregular fights doctrinally defined as "a violent struggle among state and nonstate actors for legitimacy and influence over the relevant population."[1] However, for most of the military establishment, irregular warfare was a secondary concern as the military-industrial complex focused its efforts and resources on the problem of defeating the conventional armed forces of the Soviet Union and other near-peer competitors. Most thought leaders in the military and in policymaking circles continued to consider irregular warfare as an aberration—an inconvenient deviation from the norm of joint warfare. As the superpower conflict of the Cold War gave way to the complex and chaotic state and nonstate conflict of the post-Cold War era, irregular warfare has become the norm. It is such a certain feature of every modern conflict that doctrine and concepts no longer treat it as secondary or exceptional. Instead, it is simply the reality of modern conflict today.

At the heart of resistance are people who choose to oppose government authority. Resistance begins in the human heart and later expresses itself through protests, demonstrations, strikes, clandestine organizations, underground newspapers, sabotage, subversion, guerrilla warfare, and eventually civil war. The problem becomes an urgent matter for strategists and military leaders when resisters organize and transition from nonviolent protest into militarization. However, to understand the phenomenon, it is crucial to look beyond the masked guerrilla and grasp at the nature of resistance itself—its causes, dynamics, and evolution. Instead of dealing with resistance as a secondary form of warfare, the goal is to take it on as a subject in its own right.

The study of resistance has continued to evolve throughout the twentieth century and into the twenty-first century as policymakers, defense officials, and social scientists in academia have sought to answer similar questions: What is resistance? How does one know if a country is experiencing resistance? What are the most important characteristics of resistance? The answers to these questions have shifted considerably from the early Second World War era into the post-Cold War period alongside changes in the international system, trends toward globalization, advances in transportation and communication technologies, the availability of resources needed for mobilization, and changes in international norm regimes, among other conditions. One of the most striking characteristics is the dramatic drop in traditional interstate wars, although armed competition among states still occurs, but in more nuanced forms.

The period since the Second World War has also seen tremendous changes in the kinds of resistance movements that mount campaigns against incumbent governments or occupying forces. Nonviolent civil resistance movements were highly effective in achieving their objectives, more so than violent resistance movements. The kinds of actors participating in all sorts of resistance movements, whether violent or nonviolent, have also evolved as nonstate actors participate in greater numbers than ever before. During this period, civil resistance movements in the Middle East ousted the so-called "enduring" authoritarian regimes with considerable security force capabilities in the Arab

Spring. Similarly, mass protests against electoral fraud ousted the president in Ukraine's 2005 Orange Revolution. These historical developments have complicated how scholars, policymakers, and military personnel think about resistance, highlighting important distinctions that the Special Forces should understand to be effective in our rapidly evolving security environment.

EARLY DEFINITIONS OF RESISTANCE

The modern use of the word *resistance* was first based on the domestic insurgent movements of Europe against Axis occupying powers in World War Two, especially the French Resistance.[2] The conception of resistance as a domestic effort against an encroaching foreign force had a lasting impact on definitions of the term throughout the US military as communist regimes were understood as politically executed occupations. In the original 1963 publication of the seminal *Undergrounds in Insurgent, Revolutionary, and Resistance Warfare*, the Special Operations Research Office (SORO) cited a 1949 text, defining resistance as distinct from revolutions and insurgencies: "operations directed against an enemy, behind his lines, by discontented elements among the enemy or enemy occupied population."[3] This conception of resistance as inherently insurgent in nature remains influential. The 2012 *Guide to the Analysis of Insurgency* defines resistance as a type of insurgency (distinct from revolutionary, separatist, and other variants) that "seek[s] to compel an occupying power to withdraw from a given territory."[4] The definitions rely on describing resistance as corollary of interstate war and as a product resulting from armed violence.

Alternatively, there has also been a broadening strain of thought on the nature of resistance in two regards: first, that resistance does not need to be against a foreign occupier, and second, that it does not need to be carried out through the predominant use of violent tactics. While composing the second volume of SORO's *Casebook on Insurgency and Revolutionary Warfare* in 2012, the Assessing Revolutionary and Insurgent Strategies (ARIS) team, under the direction of the United States Army Special Operations Command (USASOC), outlined why it was necessary to broaden the criteria for case selection to include nonviolent instances of revolution:

The original SORO study required that a revolution had to involve the use of, or the threat of use of, force. Force was open violence, guerrilla warfare, or even civil war within the military. Mere propaganda, protest, strikes, or even passive resistance did not meet the criteria in the previous study. For modern revolutionary warfare, we have decided to include examples of revolutions begotten by either entirely or predominately nonviolent means . . . These "velvet" revolutions can teach the student of modern warfare how mass protests, coordinated propaganda and information campaigns, and coercive yet nonviolent means can topple an entrenched governmental system.[5]

The trends in US government and military assessments of resistance have culminated in the 2016 doctrinal definition, which defines a resistance movement as "an organized effort by some portion of the civil population of a country to resist the legally established government or an occupying power and to disrupt civil order and stability."[6]

RESISTANCE DEFINED IN THE ARIS PROJECT

Resistance is a difficult concept to isolate. Although many modern conceptions of resistance are not essentially violent or even political, it is important to discern between those definitions or types of resistance that are strategically relevant and those that are not. If the Special Forces maintains that only political, armed, and violent variations (insurgency, guerrilla acts, and terrorism) deserve study or attention in an area of operations, he or she will become susceptible to strategic surprise if movements or groups that have been ignored because they are considered nonviolent decide to begin using violent or even armed tactics. On the other hand, however, the Special Forces need to focus their attention on those actors of strategic significance, so a useful definition of resistance must be bounded enough to avoid distracting the Special Forces into focusing on "resistance" groups of little to no consequence to the mission. A careful balance is required between broadened situational awareness and discernment of strategic relevance.

The definition for resistance presented by the ARIS team in the 2016 *Conceptual Typology of Resistance* seeks to strike this balance. "Resistance," it states, "is a form of contention or asymmetric conflict involving participants' limited or collective mobilization of subversive and/or disruptive efforts against an authority or structure."[7] Several concepts here need to be elaborated. First, resistance is asymmetric, meaning that the conflict or contention is being waged by a less powerful actor against a more powerful or authoritative opponent. Second, it must involve the mobilization of people (i.e., a single person is not a resistance unto themselves), whether it be a mass or a smaller conspiracy. Third, resistance activities must be at least subversive, but they are also primarily disruptive. This means that the resistance must at least be planning and acting to undermine the power and authority of its opponent, even if its operations are not widely apparent. When operations do emerge, they cause disruption in the normal operation and functioning of the ruling authority.

As a general rule, USASOC has operational concern only for political forms of resistance that involve armed or at least violent action. However, fewer resistance movements in the twenty-first century start as armed or violent endeavors. Civil resistance, where mass mobilization to protest and/or riot with an implicit threat of armed violence to compel a government to change policies or even step down from power, is now a feature of world politics, even causing regime change across the Middle East and in Ukraine. The importance of civil resistance is likely to continue as globalization and technology contribute to the erosion of sovereignty.

Resistance does not begin when groups pick up guns or casualty counts hit an arbitrary number. Instead, it is a continuous concept, with a range of potential tactics and objectives ranging from the discreet and nonviolent to the revolutionary and genocidal. The Special Forces, when assessing the political and human landscape of an area of operations, must seek to discern between those movements with potential to evolve into military, or at least strategic, players in the battle space, and those who will remain on the sidelines of the conflict.

The continuum of resistance includes analysis of different strategies and tactics but also the shifting legality of resistance movements.[8] Figure 1 illustrates the different categories of resistance from nonviolent, legal resistance to full-fledged belligerency that relies on the use of violence to achieve its objectives. Although it periodically shifted from legal to illegal, the Solidarity movement in Poland is an example of nonviolent legal resistance.[9] On the belligerency end of the spectrum, the Fuerzas Armadas Revolucionarias (FARC) in Colombia serves as an apt example.[10] Resistance movements transition along the continuum as the intensity and duration of the violent conflict increases as well as when a resistance movement adopts more sophisticated organizational structures. However, although this legal continuum is helpful for analysis, it is oftentimes difficult to clearly define the boundaries between each category when examining resistance movements.

Identifying shifts along the continuum is important in several regards. First, from an academic perspective, it is useful to categorize resistance movements at particular points in time to enable cross-case comparison to identify regularities or isolate factors that contribute to different outcomes in each conflict. Secondly, from an operational perspective, identifying a resistance movement's place on the continuum is crucial because it signals when international humanitarian law (IHL) applies. IHL impacts the legal status of persons in resistance, whether US personnel or indigenous members of the movement, changing how foreign governments must treat the individuals when captured to ensure the authorities are complying with international law.

It is important to recognize these objective parameters that help identify transitions along the continuum are not the only parameters. Because resistance movements are challenging incumbent political authorities, identifying the transitions is also a matter of politics or policy.

Increasing level of intensity, duration, and organization

Nonviolent Resistance

Use of Legal Processes for Political Advantage

Characteristics: Individuals or groups use legal processes to resist standing government, e.g., social media messaging, peaceful demonstration, canvassing polls.

Corresponding Legal Status: Individuals are subject to HN law. Indigenous status would be as a citizen/resident. UW status is as a tourist subject to HN jurisdiction. FID status would depend on any applicable SOFA.*

Third-Party Involvement: Foreign government support unlawful unless HN consents; discovery of the presence of personnel could prompt diplomatic problems, charges of espionage.

Illegal Political Acts

Characteristics: Individuals or groups resorting to illegal political acts to resist standing government, e.g., refusal to comply with certain laws (civil disobedience) or other disruptive, nonviolent acts.

Corresponding Legal Status: Individuals are subject to HN civil and criminal law. Indigenous status would be as a citizen/resident. UW status is as a tourist subject to HN jurisdiction. FID status depends on any applicable SOFA.* Diplomatic channels can be used to negotiate jurisdiction or release.

Third-Party Involvement: Foreign support to domestic criminals unlawful unless HN consents; discovery of support could prompt diplomatic tension, accusations of aggression, charges of espionage.

Armed Resistance

Rebellion

Characteristics: Short-term, isolated, violent engagements of low intensity by a group (e.g., riots); law enforcement mechanisms are able to suppress the violence; it remains a domestic matter.

Corresponding Legal Status: Rebels are subject to domestic criminal law. UW status is as a tourist subject to HN jurisdiction. FID stand alongside the HN government under applicable SOFA.* Diplomatic channels may be used to negotiate jurisdiction or release. NOT yet an armed conflict so IHL does not apply.

Third-Party Involvement: Support of rebels violates HN sovereignty and contravenes international norms of noninterference, with some exceptions. Support to HN by invitation or with consent is permissible.

Insurgency

Characteristics: Recognition of an insurgency is based on facts and political factors. In general, the fighting is more sustained and intense and cannot be easily suppressed by the government. Other elements include increased levels of the insurgent group's organization and territorial control.

Corresponding Legal Status: As a NIAC, the IHL protections of Common Article 3, and potentially Additional Protocol II, apply. Parties can agree to apply more protections but not fewer.

Third-Party Involvement: Support of insurgents violates HN sovereignty and contravenes norms of noninterference, with some exceptions. Support to HN via COIN permissible with consent. States may engage with insurgents to protect property, commercial interests, nationals.

Belligerency

Characteristics: (1) A general as opposed to local armed conflict, (2) belligerents administer a substantial portion of territory, (3) belligerents follow laws of war and use a command system, and (4) circumstances require states to define their positions in relation to the conflict.

Corresponding Legal Status: LOAC applies. The resistance group is deemed a de facto state if the host nation recognizes the resistance as a belligerency, and its forces receive combatant/ POW status. The resistance group and HN and allies are bound to apply LOAC, the customary law of international armed conflict.

Third-Party Involvement: Recognition of a resistance group as a de facto state imposes a duty of neutrality on third-party states. Third parties may support one side or the other, but doing so constitutes an act of war.

* SOFA or other international agreement.

COIN—Counterinsurgency
HN—Host nation
FID—Foreign internal defense
IHL—International humanitarian law
LOAC—Law of armed conflict
NIAC—Noninternational armed conflict
POW—Prisoner of war
SOFA—Status of forces agreement
UW—Unconventional warfare

Figure 1. Continuum of categories of resistance.

It is generally not in a state's interest to identify challengers as advancing on the resistance continuum, particularly where it requires treating prisoners under IHL. In many cases, it is preferable to categorize challengers as criminals, not as an organized resistance, subject only to domestic laws as identifying substantial threats to a state's sovereignty saps its legitimacy.[11, 12] However, some resistance activities considered illegal by a government, such as speech critical of government, is protected under international human rights treaties. In these circumstances, persecution of individuals for their political beliefs can be the basis for claims for asylum in other countries.

ARIS Resistance Attributes

The ARIS research team has expanded the concept of resistance to enable a firmer understanding of its key attributes.[13] Regardless of the definitions used to describe resistance, treating it as a holistic phenomenon present in nonviolent and violent manifestations allows observers to unpack the dynamics between organizations involved in collective action and their environment.

The focus is on identifying key attributes of resistance that can aid students and practitioners in preparing nuanced analyses of the resistance phenomenon. In previous ARIS works, researchers identified a set of sociological concepts, agency and structure, that provides a comprehensive framework for thinking about the dynamics of resistance. Agency and structure refer to the dialectic between human will and the surrounding environment that guides or channels it. Agency refers to the capacity of individuals to act and make their own free choices, whereas structure refers to broader forces that constrain or shape an individual's capacity to take action. The structures encompass a variety of factors, including political, social, and economic conditions but also norms or values that prescribe appropriate behavioral conduct. Structural forces also include institutional arrangements, such as governments or religious institutions, each with the capacity to constrain or otherwise shape individual behavior. The dialectic that emerges between structure and agency alerts observers to the key attributes of resistance described later: actors, causality, opportunity, organization, and actions.

Actors

Resistance requires the participation of actors, defined as those persons or groups that may be directly or indirectly involved in a resistance at any given time. This attribute broadly encompasses the individual participants in an organized resistance, formal or informal leaders of the movement, the general population (or mass base), active or passive followers or sympathizers, entrepreneurs, opportunists, and external supporters (diaspora, nongovernmental organizations, or foreign governments).

Causality

Causality refers to the expressed reason or motivation for a specific occurrence of resistance. Often presented as grievances within an ideological or narrative framework for mobilization against a resisted structure, this attribute of resistance encompasses the multidirectional dynamics between the status quo of those in power and the objectives of those who resist. Historical or ethnic dynamics may be at play in the causality of a given resistance, but personal slights, self-radicalization, and other factors must not be ignored in assessing a holistic picture of the foundational dynamics of the conflict.

The idea of accommodation is an important factor within this attribute. Although the causal objectives or grievances may exist within certain actors, the ability of the power or authority structure to accommodate and alleviate the mounting pressures of discontent that may result in resistance can have a significant impact on the emergence, intensity, or length of a given movement. Particularly, participatory means through which to express and seek recourse for grievances could diffuse a resistance by denying it a causal foundation. In this way, the causality of many resistance movements, although not all, may be characterized as a lack of accommodation to the actors.

Environment

The environment attribute captures the preexisting and emerging conditions within the political, social, or economic contexts that enable or constrain the mobilization of resistance whether directly or indirectly. Because there are many environmental factors possibly contributing to resistance mobilization, it is important to focus research efforts on those that are the most strategically significant. In the ARIS

11

typology, these factors include the characteristics of the state, social structures, the legal environment, and geography, among others.

Organization

Every resistance has an organizing principle or idea around which actors coalesce, as well as some structure (formal or informal) through which actors conduct resistance. The rallying points for resistance can include a shared identity (e.g., cultural, ethnic, or religious), ideological imperatives (e.g., freedom and justice), a common objective (e.g., secure power or basic necessities), charismatic leadership, and/or other incentives. In terms of structure, the organization of resistance can range from broad-based, decentralized, and leaderless networks to rigidly formed hierarchical structures. The organizing principle and structure can change over time, evolving in reaction to the shifting landscape of opportunities. Other concepts and behaviors associated with the attribute of organization include recruitment, fundraising, security, and centrifugal versus centripetal dynamics. Organization as an attribute preserves the essentially collective nature of resistance as a phenomenon.

Action

Actions are the means by which actors carry out resistance, engaging in behaviors and activities in opposition to a resisted structure. Initiated by the agency of individuals organized in pursuit of common goals, actions can encompass both the specific tactics used by a resistance movement and the broader repertoire for action (i.e., strategy). Associated concepts include both violent tactics (terror, targeted violence, sabotage, etc.) and nonviolent tactics (protest, propaganda, etc.), as well as means for external actor involvement (unconventional warfare) and the decisions to strategically restrict actions (legality, targeted versus indiscriminate, etc.) according to opportunities and the need for legitimacy.

VIOLENT AND NONVIOLENT RESISTANCE

· The study of resistance as a general phenomenon observable across a spectrum of cases ranging from nonviolent to violent forms is complicated by decades of research focused almost exclusively on violent resistance. The trend has begun to change as scholars have made considerable conceptual shifts in their studies of resistance and gathered enough data on nonviolent campaigns to be able to make worthwhile comparisons between violent and nonviolent campaigns. The evidence from this emerging field of study revealed surprising findings. Nonviolent campaigns increased in frequency over the past one hundred years and were more effective in reaching a group's political objectives than violent campaigns.

The most prominent exception to the exclusive study of violent resistance campaigns is the large body of research on social movements. Political scientists, sociologists, historians, and others in the field studied the mass movements that relied on civil resistance campaigns against incumbent regimes to better understand how the groups formed and mobilized and under which conditions they succeeded or failed. Researchers identified important theories, concepts, and mechanisms that developed into a robust research program that helped to explain the variation in outcomes among social movement cases. One important shortcoming of the field, however, was its overriding interest in explaining social movements struggling for goals that coincided with Western liberal values, such as civil rights, labor movements, or antiwar movements. Not only were actors that did not necessarily align with these values also forming successful social movements, such as political Islamic movements in the Middle East, but sometimes nonviolent campaigns morphed into violent insurgencies, such as the decades-long conflict in Northern Ireland.

As a result, others began to view resistance holistically, arguing that the same theories, concepts, and mechanisms used in social movement theory could also usefully be applied to explore research questions regarding violent campaigns. The research program, called contentious politics, argues that resistance occurs on a spectrum from nonviolent campaigns to full-blown insurgencies. Most importantly, the overriding theory of contentious politics is that the incidents, whether social movements, revolutions, peasant rebellions, or mass protests, are all different manifestations of the same category of phenomena. The theory of contentious politics is still just that—a theory—and scholars are only beginning to empirically test whether or not violent campaigns are a wholly different animal from nonviolent campaigns.

The contentious politics research program, in turn, coincided with other developments in research on political violence. Most research on war, outside of the research on interstate war in international relations, has focused on domestic intrastate war, or civil war, because this is the predominant form of conflict today. Research on civil war has largely treated political conflict as synonymous with violence. In practice, that means that civil wars are coded dichotomously. A "0" indicates that in a particular case, usually a country or a year, violence is absent or at a very low level. On the other hand, a "1" means that violence is present. A "1," then, is a positive case of civil war, whereas a "0" indicates the case is negative, that is, that there is no civil war.

While the method is certainly intuitively simple, it may have prejudiced some researchers to think about conflict strictly in terms of violence. The absence of violence is equated with the absence of conflict. In reality, however, the absence of violence in a country can mask any possible number of scenarios. It is certainly possible that there really is no ongoing conflict in society. On the other hand, it could mean that there are ongoing conflict processes in the country, such as massive, widespread protests, boycotts, demonstrations, or other forms of nonviolent resistance coordinated by organized actors. In this latter case, the country might be on the cusp of violence as the state violently represses dissenters, leading them to transition to predominantly violent tactics. Of course, such a trend was evident in Syria during the Arab Spring. Several months before the Free Syrian Army and the Assad regime began shooting at one another, dissenters took cues from events occurring elsewhere in the region during the Arab Spring and began organizing and mounting their own civil resistance against the authoritarian Assad regime. The resistance, or conflict processes, in Syria began long before the shooting actually started.

Political Science

Political science views resistance as a kind of domestic political conflict. Political conflicts as a whole can be violent or nonviolent, and the actors involved can include individuals, states, international organizations, and nonstate groups. There are four essential components for consideration as a political conflict: (1) the conflict is over "values relevant to society," also called conflict items;[a] (2) the conflict is between at least two decisive and directly involved actors; (3) the conflict is carried out by actions (also called conflict measures) "that lie outside established regulatory procedures;" and (4) the conflict must "threaten core state functions, the international order or hold out the prospect to do so." Domestic violent and nonviolent political conflict (i.e., resistance) is studied in political science as civil wars[b, 14] insurgencies, political violence, rebellion, nonviolent uprisings, domestic strife, state failure, and political instability. Political science conducts the lion's share of research on the phenomenon of resistance, particularly violent resistance, but has adopted theoretical contributions from other disciplines such as social movement theory from sociology.

[a] Conflict items refer to the justifications for fighting and the grievances held for by actors, most commonly issues such as territory, secession, decolonization, autonomy, system/ideology, national power, regional predominance, international power, and resources. Political conflicts can also involve "ideological, religious, [and] socioeconomic" factors as conflict items, but the actors must seek a change in the political system for the conflict to be considered genuinely political (Heidelberg Institute for International Conflict).

[b] Political science research on civil wars as laid out by James Fearon includes coups d'état, popular revolutions, "classical" civil wars, secessionist wars, ethnic wars, ideological wars, wars of decolonization, and peripheral insurgencies.

Overcoming the assumption that only violence is equated with conflict has led to a conceptual shift in the study of conflict processes. One conceptual shift has already been described; violence may or may not be wholly distinctive from nonviolent conflict. If it is, then political violence emerges by itself without any attendant nonviolent conflict. However, it is more likely that violence is another step in existing, unresolved, nonviolent conflicts. From this latter perspective, it is possible to identify the conditions under which actors transition to violence from nonviolence.

The assumption that all resistance is part of the same phenomena is that the adoption of violence is a choice made by actors. Currently, due to the theoretical limitations previously mentioned, researchers have not paid sufficient attention to the conditions or triggers that prompt actors to adopt violence at a particular time and place. In the case of the Northern Ireland conflict, the PIRA decided explicitly to adopt a predominantly armed campaign in 1969. The Republicans, although in a different organizational form, relied on nonviolent civil resistance to pressure the British government to withdraw from Northern Ireland for most of the previous decade. Why 1969? To answer this crucial question, it is necessary to look at a range of conflict cases that utilize violence and others that do not. Moreover, it is also necessary to gain a better understanding of the full range of possible options available to the actors, including various nonviolent and violent strategies. Research in this area is just beginning; where appropriate, the initial findings of this research are discussed in chapter 5.

Another related assumption regards the presumed effectiveness of violence. Actors are assumed to use violence to press their claims simply because it is the most effective means at their disposal to wage political struggle. The implicit theory is that those making the decision are rational actors, carefully estimating the costs and benefit ratio of armed violence. When the perceived benefits outweigh the costs, then the actor adopts violent tactics. However, the assumption that actors turn to violence because of its effectiveness appears to be flawed. In some cases, violence is actually a suboptimal outcome in which actors are less likely to achieve their goals once violence enters the equation. Certainly, violence is costly in terms of commitment and resources, although it may be that the actors who employ violence have access to only the resources and strategies that facilitate violence.

Recent work on nonviolent resistance demonstrates that it can be an effective strategy for helping actors to achieve their political objectives.[15] Previously, most research on nonviolent campaigns used qualitative case study methods that are rich in detail. The research propelled the social movement and contentious research programs, helping to develop the theories, concepts, and mechanisms necessary to enable researchers to gain a better understanding of nonviolent campaigns that emerged and mobilized and the conditions under which the campaigns succeeded or failed.[16]

The compilation of data on nonviolent campaigns has increased the possible scope of research on nonviolent campaigns. In 2011, a group of researchers released a dataset that included nonviolent and violent resistance campaigns from 1900 to 2011.[17] The arduous task has made possible, for the first time, research that compares the outcomes of violent versus nonviolent campaigns. The data includes 323 cases, including one hundred nonviolent campaigns since 1900.[18, 19] Research using the dataset determined that, overall, nonviolent campaigns are twice as likely to achieve their political goals than violent campaigns.

The finding holds even when researchers control for factors likely to impact those results, including the target regime type and the target regime capabilities. However, others have contested this finding, arguing that nonviolent strategies work best in "easy" cases, where the regime is considered more likely to acquiesce to demands by resistance movements. In some cases, violent strategies may be adopted in more difficult cases where the government is likely to stand resolute in the presence of protest.

Sociology

Sociology views resistance as a type of social movement. Social movement theory is a multidisciplinary field across the social sciences, descending from the historical study of popular revolutions. Although there is no authoritative definition of social movements, it is generally agreed that social movements have three essential characteristics: (1) an informal network between a plurality of individuals, organizations, and/or groups (2) who are engaged in cultural or political conflict (3) based on a shared collective identity. The focus of social movement theory has largely moved to "new social movements" (i.e., those since the 1960s that have focused on identity and quality of life), as opposed to those that seek specific public policy or economic changes. Resistance has been defined very broadly in sociology, causing an ambiguous academic environment, where *action* and *opposition* are the core elements of resistance according to sociology, but the sense of action is very broadly conceived. Although many of the "resistance" movements studied in social movement theory have little to no overlap with those of interest to the Special Forces, their study still offers key insights into how resistance behaves as a phenomenon of collective human behavior.

Conflict Barometer 2013 (Heidelberg, Germany: Heidelberg Institute for International Conflict, 2014), http://hiik.de/de/downloads/data/downloads_2013/ConflictBarometer2013.pdf; Crane Brinton, The Anatomy of Revolution, Revised Edition (New York, NY: Vintage Books, 1965); Lyford P. Edwards, *The Natural History of Revolutions* (Chicago, IL: University of Chicago Press, 1927); Paul Meadows, "Sequence in Revolution," *American Sociological Review* 6, no. 5 (October, 1941): 707–709; Rex D. Hopper, "The Revolutionary Process: A Frame of Reference for the Study of Revolutionary Movements," *Social Forces* 28, no. 3 (1950 March): 271–279; Mario Diani, "The Concept of Social Movement," *The Sociological Review* 40, no. 1 (1992): 1–25; Jonathan Christiansen, *Social Movements & Collective Behavior: Four Stages of Social Movements* (Ipswich, MA: EBSCO Publishing, 2009); Jocelyn A. Hollander and Rachel L. Einwohner, "Conceptualizing Resistance," *Sociological Forum* 19, no. 4 (2004): 533–554.

Researchers speculate that nonviolent campaigns may have an advantage because of the mass-based participation that characterizes the successful campaigns. Violent campaigns exhibit participation problems because violence requires a higher level of moral, physical, and informational commitments than required in nonviolent campaigns. Not everyone is suited to perpetrating acts of violence, but considerably more people are suitable for participating in mass protests, sit-ins, or boycotts. Mass-based participation has advantages in several key areas: "enhanced resilience, higher probabilities of tactical innovation; expanded civic disruption (thereby raising the costs to the regime in maintaining its status quo), and loyalty shifts involving the opponent's erstwhile supporters, including members of the security forces."[20]

ROAD MAP

This ARIS primer is intended as an introduction to the latest science of resistance for advanced students of resistance in the Special Forces. The theories, concepts, and evidence discussed in the following chapters draw from all areas of the social sciences but particularly political science and sociology, the two disciplines that have produced the most research on resistance. Some of the research presented deals with violent and nonviolent resistance separately, while other research looks at resistance holistically.[21] The ARIS project aims to extend this latter subfield of research to advance the science of resistance past its current preoccupation with primarily violent resistance. This primer also incorporates work conducted in previous ARIS volumes, including *Human Factors Considerations of Undergrounds in Insurgencies* and *Undergrounds in Insurgent, Revolutionary, and Resistance Warfare*. Where appropriate, it draws illustrative examples from the *Casebook on Insurgency and Revolutionary Warfare, Volume II: 1962-2009* and other ARIS publications.

The chapters in this primer introduce large, overriding questions in studies of resistance. Each question has engaged many scholars of resistance, and attempts to answer these questions have resulted in large bodies of research. In describing the latest research that seeks to answer these questions, the discussions focus on the key attributes of resistance discussed earlier, describing how resistance actors interact with their surrounding environment, from the political opportunities they encounter to the organizational strategies they adopt.

In chapter 2, the analysis centers on why some countries experience conflict, while others do not. The discussion is based on research related to civil war onset in political science but also literature from social movement theory and the contentious politics research program that assesses how interactions between structure and agency impact the decisions and actions of resistance movements. The chapter first introduces structural variables, including measures of geography, economic grievances, ethnic grievances, and regime types and governance.

These variables capture the macro, large-scale institutions, conditions, or patterns within which organizations and individuals act. Beyond the structural variables, the latter part of this chapter introduces the theory of political opportunity developed in the contentious politics research program. Political opportunity is a concept that refers to the interactions between resistance actors and the environment in which they operate, explaining how conditions in the surrounding political environment incentivize individuals or groups to make decisions or take actions that they otherwise might not have been able to had those opportunities not been present.

One of the most enduring puzzles in resistance studies regards what leads some people or groups to mobilize to collective action. The question is an enduring one because, from the observer's perspective, mobilization is a difficult endeavor. While grievances are plentiful, resistance movements are comparatively rare. Chapter 3 introduces the predominant theories on mobilization that inform this perspective, including the free-riding problem of collective action and how selective incentives can overcome the inherent difficulties of collective action. At the cognitive level, psychological risk factors sometimes play a role in mobilization, although less so than conventional wisdom suggests. Other analysis looks at organizational factors that contribute to mobilization, including the role of social networks and affiliative factors that encourage individuals to reduce uncertainty through group collective identity. Finally, the chapter introduces micromobilization processes in the control-collaboration model that assess how interactions between civilians and armed actors influence mobilization in unexpected ways. The seemingly divergent theories and processes discussed in the chapter are holistically reviewed in terms of resistance movement phases developed in previous ARIS works.

Chapter 4 addresses one of the pressing questions motivating the ARIS project: why do some groups adopt some strategies and tactics over others? While some resistance movements rely on predominantly nonviolent tactics, other movements use violence exclusively to press their political claims or a mixture of violent and nonviolent strategies. Most scholars explain these decisions, particularly those related to the use of violence, as resulting from the movement's rational calculations regarding its expected costs and payoffs. However, other approaches to the question have revealed how violence emerges from other factors, including interactions with other actors inside and outside the movement. The interactions prompt mechanisms related to competition over scarce resources that can propel a movement to use violence. This chapter explores not only the transition to violence but also why groups use certain forms of violence over others, such as indiscriminate or selective violence. It assesses how organizational structure, resourcing, and territorial control impact a group's use of violence; the lethality, target, and scale of violence; and interactions with the host population.

While the previously described chapter investigates the motivations of resistance, chapter 5 addresses the equally important question of how resistance movements terminate or demobilize and the stability of the postconflict environment. Like previous chapters, it includes a mixture of research on the specifically violent resistance characterizing civil wars, but it also addresses the rich theoretical and empirical material of social movement theory that incorporates both violent and nonviolent resistance. The chapter examines conflict termination but also the risk that a country or region will fall back into war after a conflict has ended, called recidivism. Violent domestic conflicts can end as rebel victories, government victories, or negotiated settlements. In the post-Cold War era, negotiated settlements are especially prevalent as third parties help opponents reach a political solution to the violence. However, settlements are particularly vulnerable to recidivism because of the lack of trust and commitment between the parties to the conflict. Finally, the chapter concludes with a review of the dynamics and theories related to the decline of social movements and nonviolent resistance as laid out in social movement theory and the field of contentious politics. Demobilization is discussed in terms of success and failure as movements are co-opted by the incumbent regime, establish themselves within the mainstream political system, or fade out because of exhaustion, abeyance, or repressive measures by state security forces.

ENDNOTES

[1] US Department of Defense, Joint Publication 1, *Doctrine for the Armed Forces of the United States* (Washington, DC: CreateSpace Independent Publishing Platform, 2013), 1–6.

[2] W. Sam Lauber, Steven Babin, Katherine Burnett, Jonathon Cosgrove, Theodore Plettner, and Catherine Kane, *Understanding States of Resistance (Fort Bragg, NC: US Army Special Operations Command (USASOC)*, forthcoming).

3   Julien Amery, "Of Resistance," in *The Nineteenth Century Magazine* (March 1949): 138–149; as quoted in Andrew R. Molner, William A. Lybrand, Lorna Hahn, James L. Kirkman, and Peter B. Riddleberger, *Undergrounds in Insurgent, Revolutionary, and Resistance Warfare* (Washington DC: Special Operations Research Office, 1963).

4   US Government, *Guide to the Analysis of Insurgency* (Washington, DC: United States. Central Intelligence Agency, 2012), 3.

5   Chuck Crossett, ed., *ARIS Casebook on Insurgency and Revolutionary Warfare, Vol II: 1962–2009* (Fort Bragg, NC: USASOC, 2012), xv–xvi.

6   US Joint Chiefs of Staff, *Joint Publication 1-02, Department of Defense Dictionary of Military and Associated Terms* (Washington DC: Joint Chiefs of Staff, 15 February 2016), 204.

7   Jonathan B. Cosgrove, and Erin N. Hahn, *ARIS Conceptual Typology of Resistance* (Fort Bragg, NC: USASOC, in press), 9.

8   Erin Hahn, ed., *ARIS Legal Implications of the Status of Persons in Resistance* (Fort Bragg, NC: USASOC, n.d.), 1–16.

9   See Chuck Crossett and Summer Newton, "Solidarity," in *ARIS Casebook on Insurgency and Revolutionary Warfare*, Chuck Crossett, ed., 645–670. (Fort Bragg, NC: USASOC, 2013).

10 See Ron Buikema and Matt Burger, "Fuerzas Armadas Revolucionarias De Colombia (FARC)," in *ARIS Casebook on Insurgency and Revolutionary Warfare*, Chuck Crossett (ed.), 31–54. (Fort Bragg, NC: USASOC, 2013).

11 The status of prisoners, for instance, was a particularly contentious issue in Northern Ireland during the height of the conflict between the Provisional Irish Republican Army (PIRA) and Northern Irish and British security forces. In 1975, the British eliminated the special category status for PIRA prisoners to delegitimize the group. The policy shift modified their status from political prisoners to criminals. In response, PIRA prisoners launched an organized prison protest known as the "blanket" protest as the prisoners refused to don the orange uniforms worn by criminal inmates, instead wearing only a blanket. The protest, although not successful in changing British policy, was highly effective in gaining popular support.

12 See Chuck Crossett and Summer Newton, "The Provisional Irish Republican Army (PIRA): 1969–2001," in *ARIS Casebook on Insurgency and Revolutionary Warfare*, Chuck Crossett, ed., 315–318. (Fort Bragg, NC: USASOC, 2013).

13 Cosgrove and Hahn, *ARIS Conceptual Typology of Resistance.*

14 James D. Fearon, "Why Do Some Civil Wars Last So Much Longer Than Others?" *Journal of Peace Research* 41, no. 3 (2004): 275–301.

15 Erica Chenoweth and Maria J. Stephan, *Why Civil Resistance Works: The Strategic Logic of Nonviolent Conflict* (New York, NY: Columbia University Press, 2011), 3–30.

16 Increasingly, the research is also used to facilitate a better understanding of violent campaigns.

17 The Nonviolent and Violent Campaigns and Outcomes (NAVCO) dataset is available at the NAVCO Data Project website, http://www.du.edu/korbel/sie/research/chenow navco_data.html, accessed on March 23, 2016.

18 The campaigns included in the dataset represent all known unarmed and armed cases between 1945 and 2006 in which resistance actors held maximalist goals of overthrowing the existing regime, expelling foreign occupations, or achieving self-determination at any point during the cyle of the resistance campaign.

19 Erica Chenoweth and Orion A. Lewis, "Unpacking Violent Campaigns: Introducing the NAVCO 2.0 Dataset," *Journal of Peace Research* 50, no. 3 (2013): 415–423; Chenoweth and Stephan, *Why Civil Resistance Works*, 17.

20 Chenoweth and Stephan, *Why Civil Resistance Works*, 10.

21 As appropriate, readers are alerted to the scope of the research being presented.

CHAPTER 2.
WHY DOES RESISTANCE OCCUR IN SOME COUNTRIES OR REGIONS?

Why do violent insurgencies appear in some countries and persist for decades, while countries that appear culturally, politically, or economically similar experience no such events?[1] Many people would like to be able to predict when and where such violence occurs. For decades, social scientists have been trying to understand which factors increase the probability of terrorism, insurgency, or other forms of political violence occurring in a state. While there is no universal law capable of predicting the onset of resistance, the available research points to factors that make some states more susceptible to political violence than others.

This chapter reviews the conditions that increase the risk that a country or place will experience internal armed conflict. Structural variables are the first set of conditions introduced in this chapter. These variables capture the macro, large-scale institutions, conditions, or patterns within which organizations and individuals act. Insurgencies form and operate in areas shaped by geographic, social, economic, and political systems. These features impact decisions and behaviors surrounding a group's formation, longevity, strategies, and chances for success.

The focus on structural variables has its limitations, particularly for Special Forces, who need to translate research into actionable data. Although we know that mountainous terrain is a geographic variable that increases the likelihood of violent conflict, it is a permanent feature about which little can be done. The research on structural variables typically excludes more fine-grained analysis that looks closely at the behavior of actors involved in the conflict that contributes to resistance or violence. Structural variables, particularly when they are fixed (such as terrain), do not shed much light on the triggers that lead directly to the adoption of violence by one or more actors in a conflict.[2]

As a result, the latter part of this chapter moves beyond structural variables to a theory developed in the contentious politics research program. The theory is based on the concept of political opportunity. It is especially helpful because it helps observers identify the more immediate triggers to violence as it incorporates elements of structure and agency. Structures refer to the overarching patterns of the physical or social world that impact individual behavior in recognizable, predictable ways. Agency, by contrast, is a concept that refers to the capacity of individuals to act in accordance with their own preferences or beliefs. When structure is overemphasized, it makes people appear as if they are mindless automatons responding only to stimuli in their external environment. However, if agency is overemphasized, it makes actors appear as if they are acting in a vacuum with no constraints from the outside world. Both positions are problematic in isolation for explaining why and how resistance movements emerge.

One of the advantages of the contentious politics research program is that the theories, concepts, and mechanisms find a balance between structure and agency. As one scholar in this field notes, "The wisdom, creativity, and outcomes of activists' choices—their *agency*—can only be understood and evaluated by looking at the political context and the rules of the game in which those choices are made—that is, *structure*."[3] Emphasizing conflict processes as an interaction between structure and agency is particularly important for the Special Forces because the value of social science research is ideally the ability to translate academic findings into best practices for resistance operations.

Nevertheless, the research on structural conditions is important because it offers guideposts to those interested in determining which countries or regions are more likely to experience conflict than others. One of the most robust findings in this literature, for instance, is that countries with low levels of income are more susceptible to conflict than their wealthier peers. Taken together, the structural factors are called the "correlates of war" because wherever political violence is present, some of these factors are likely involved. Most of the research on these variables relies on quantitative methods to make causal inferences. The quantitative method is usually some form of multivariate regression, which is a statistical technique that allows researchers to propose a set of predictor variables (e.g., infant mortality, unemployment, civil wars in neighboring countries, or climate) and an outcome variable (civil war outbreak) and then determine which of the variables are most strongly associated with the violent events while controlling for the other variables.

There is a large body of quantitative research studying structural factors related to civil war.[4] Most of the research relies on cross-national, time series analysis using several well-known datasets.[5] In these studies, the outcome variable (the phenomena being measured) is usually the beginning of civil war or political violence. This means that for each year, every country is coded as a "1" if there is violent political conflict in the country and as a "0" if there is not. In one of the most frequently used datasets, the Uppsala Conflict Data Program/Peace Research

Institute, Oslo (UCDP/PRIO) Armed Conflict Dataset, countries that experience an annual rate of twenty-five or more battle deaths between the government and nonstate actors receive the "1" coding.[6] Researchers using the data often distinguish between "major" conflicts and "minor" conflicts. Major conflicts are those resulting in more than one thousand battle deaths annually. The measure of one thousand battle deaths is also frequently used to indicate a full-blown civil war. Finding appropriate measures, and data, for explanatory variables can be even more challenging. For instance, for measures of poor economic development, researchers rely on a country's gross domestic product (GDP), annual household expenditures, and even nightlight emissions. Sometimes, the results of studies vary according to which measure and dataset are used.[7]

Statistical studies, while offering powerful analyses, present problems in interpretation. Sometimes, even if a study uncovers a causal relationship or mere correlation between the variables, it is not always clear what sort of underlying mechanism is driving the relationship. For instance, there may be a strong statistical relationship between low levels of income and civil war, but does this mean that people revolt simply because they are angry about being poor? Does it mean that the high levels of unemployment make it easier for insurgents to recruit members? Or, do low levels of income measure a hidden factor, such as state weakness, not even accounted for in the original analysis? Or is the relationship reversed, with political violence leading to increased poverty? This chapter discusses not only the correlations between civil war outbreak and structural conditions but also the competing theories behind the relationship.

The first part of the chapter discusses five broad sets of factors thought to contribute to political violence. Several are human factors, including economic grievances, ethnic grievances, poor governance, and state weakness. Other factors include a country's geographic features, demographic trends, regional stability, and conflict history. Each of these factors has been identified in at least one large-scale study relating it to political instability, political violence, or civil war. However, for each of these factors, one can find at least one study that does not find a statistical relationship between that factor and political violence. We have chosen to err on the side of inclusiveness to present the broadest set of possible factors.

The remainder of the chapter reviews theories that more deftly captures the interplay between actors and their environment. Political opportunity is a concept that refers to the interactions between resistance actors and the environment in which they operate. The underlying theory of the political opportunity concept is that factors in the political environment incentivize individuals or groups to make decisions or take actions that they otherwise might not have had those opportunities not been present. Generally, the theory is used to explain different processes in resistance, but most particularly mobilization, which is discussed in greater detail in chapter 3. However, other research looks at how political opportunities shape a resistance group's strategies, tactics, and claims, discussed in chapter 5.

SOCIOECONOMIC GRIEVANCES

Ethnic Grievances

By some estimates, in about 64 percent of all civil wars since 1946, opposing sides have divided along ethnic lines.[8] One of the strongest risk factors for ethnically motivated political violence is the marginalization or persecution of ethnic identity groups within a state.[26] While ethnicity is frequently correlated with conflict, it is less clear why ethnicity per se drives some conflicts and how such conflicts might differ from conflicts motivated by nonethnic grievances.

Some researchers have asked whether ethnically diverse countries are, in general, more prone to civil violence. Diversity is measured with the ethno-linguistic fractionalization (ELF) index,[9, 10] which measures the chance that two random individuals in a society would speak a different language.[11] However, ethnic diversity itself is not a risk factor. Many countries include different religious and ethnic groups that live together peacefully.[12]

However, certain ethnic structures in a society seem to predispose countries to violence. The ethnic polarization model, which corrects the ELF index, argues that an increase in the number of groups in a society decreases the chance for violence. Although somewhat unintuitive, the assumption behind the theory is that violence is less likely among highly homogenous and highly heterogeneous societies. According to this model, conflict is more likely when there are fewer ethnic groups that include a majority group and peripheral minority groups. Ethnic polarization appears to account for not only civil war onset but also a conflict of longer duration.

The ethnic polarization model explains ethnic violence as resulting from a majority and peripheral minority ethnic group configuration.

Jose G. Montalvo and Marta Reynal-Querol, "Ethnic Polarization and the Duration of Civil Wars," *Economics of Governance* 11, no. 2 (2010): 123–143; Marta Reynal-Querol, "Ethnicity, Political Systems, and Civil Wars," *Journal of Conflict Resolution* 46, no. 1 (2002): 29–54.

Ethnic identity is a difficult concept to define, but most agree that while it may appear fixed, there is actually a good deal of choice and negotiation involved. This means that analyses of ethnic conflict should be attentive to the ways in which ethnicity is shaped and reshaped throughout conflict processes. One of the classic definitions of ethnicity comes from Donald Horowitz. He argues that ethnic identity exists along a continuum of ways in which "people organize and categorize themselves," containing qualities of both birth and choice.[13] Oftentimes people are born into certain ethnic groups, but choice remains an important factor shaping identity as well. Within these parameters, Horowitz describes ethnicity as shared characteristics of appearance, language, religion, traditions, or some mixture of all of these attributes. His emphasis on the choice component of ethnic identity is an important one because it highlights how ethnicity is more fluid than it is typically presented. As a result, depending on the group's interactions with others, its language, religion, or other cultural markers might become especially important signifiers of its particular sense of groupness.[14] Nevertheless, ethnicity is a powerful interpretive and mobilizing frame among observers and political leaders alike.[15, 16]

Rwanda's Tutsi Genocide: Elite Manipulation of Ethnic Identity

In 1994, an orchestrated genocide against the Tutsi ethnic group in Rwanda led to the deaths of an estimated eight hundred thousand in a short period of time. The perpetrators, the Hutu ethnic group, which controlled the Rwandan government at the time, extensively manipulated ethnic identity to mobilize the Hutu population to participate in the killings. The divisions between the two groups mostly derive from Belgian colonial rule. The colonial powers favored the minority Tutsis, leading to generations of political and socioeconomic disparity between the two groups. The policy was based on perceived distinctions between the Hutu and Tutsi groups, arguing that the Tutsis were ethnically superior. After gaining independence, the

Hutu took control of the Rwandan government from the Tutsi monarchy. In the decades that followed, the Rwandan government battled various Tutsi insurgent groups, such as the Rwandan Patriotic Front (RFP), leading to sporadic civil wars. Hutu President Habyarimana deliberately used divisions between the Hutu and Tutsi groups to consolidate his own power after losing legitimacy because of his concessions to the Tutsi in French-led peace negotiations. Relations between the two groups rapidly deteriorated after the assassination of Habyarimana. The genocide began almost immediately after the assassination, pushed along by Hutu extremists. Communications sponsored by the Habyarimana administration and the extremists played up an impending threat of Tutsi attacks, encouraging preemptive strikes against them. Moreover, radio communications helped to dehumanize Tutsis among Hutu listeners, calling the group "alien" and "evil incarnate." As the genocide unfolded, Radio Rwanda broadcast twenty-four hours each day, encouraging its listeners to kill Tutsis, even providing tactical instruction on how best to kill their neighbors.

Bryan Gervais, "Hutu-Tutsi Genocides," in *Casebook on Insurgency and Revolutionary Warfare, Volume II: 1962–2009* ed. Chuck Crossett (Fort Bragg, NC: USASOC, 2012), 237–264.

Another structure vulnerable to armed conflict is ethnic minority rule. In this configuration, an ethnic minority rules over a different ethnic majority, excluding the majority from political power. As Ernest Gellner describes, "if the rulers of the political unit belong to a nation other than that of the majority of the ruled, this, for nationalists, constitutes a quite outstandingly intolerable breach of political propriety."[17, 18] The ethnic power relations (EPR) dataset captures the extent of political inclusion or exclusion of ethnic groups in a state. In this regard, it is more theoretically robust than its predecessor, the ELF index, which did not measure political exclusion.[19] Under these structural conditions of ethnic minority rule, violence is more likely.[20]

Ethnic minority rule occurs when a minority ethnic group rules a majority ethnic group(s) through political exclusion.

Andreas Wimmer, Lars-Erik Cederman, and Brian Min, "Ethnic Politics and Armed Conflict: A Configurational Analysis of a New Global Data Set," *American Sociological Review* 74, no. 2 (2009): 316–337.

However, ethnic grievances can also derive from economic marginalization. The theory of horizontal inequality argues that ethnic groups that experience both systemic political and economic exclusion as a group (in comparison with other groups in society) are more likely to engage in armed rebellion than others.[21, 22] Perceptions of inequality are just as important, if not more important, than the realities of inequality. In this regard, the theory is more about the gap between people's expectations and the possibilities afforded them in society. Horizontal inequality is similar to relative deprivation, but the former looks at group-level inequality while the latter looks at individual levels of inequality.[23, 24] Research conducted using this theoretical model combines measurements to capture inequality and ethnic settlement patterns by using geocoded inequality and ethnic settlement data.[25]

Horizontal inequality argues that political violence is more likely when groups experience systemic economic and political exclusion.

Frances Stewart, *Horizontal Inequalities and Conflict*, ed. Frances Stewart, (Palgrave Macmillan: New York, 2008); Lars-Erik Cederman, Nils B. Weidmann, and Nils-Christian Bormann, "Triangulating Horizontal Inequality: Toward Improved Conflict Analysis," *Journal of Peace Research* 52, no. 6(2015): 806–821; Lars Erik Cederman, Kristian Skrede Gleditsch, and Halvard Buhaug, *Inequality, Grievances, and Civil War* (New York, NY: Cambridge University Press, 2013).

Economic Grievances

Political violence is more likely to occur in countries with lower levels of economic development and less likely to occur in prosperous countries, making economic deprivation a risk factor for conflict processes. A country's overall level of economic development is an important factor influencing levels of political stability.[26] Economic development is a broad topic open to numerous interpretations in research. Researchers use different proxies to capture a country's level of economic development, sometimes alone or in combination with others.[27] Each of these has been shown to have a relationship with political violence: energy usage per capita, per-capita income,[28] infant mortality,[29] and level of secondary schooling among males.[30]

As with many of the structural variables discussed in this chapter, it is difficult to precisely pinpoint how lower levels of economic development are connected to conflict processes.[31] Researchers, and the evidence, disagree on the causal link between poverty and political violence. The most intuitive argument is likely wrong. Political conflict, particularly violent political conflict, does not result only from a population's anger about poverty or deprivations related to poor education, health care, or employment. Grievances may spur participation in food riots and petty crime, for instance, but sporadic, spontaneous outbursts are different from organized resistance movements that persist over time.

> *Economic opportunity is used to explain mobilization into resistance groups. When economic development in an area is poor, people are more likely to join a movement because it provides the best economic payoff available.*

David Keen, "The Economic Functions of Violence In Civil Wars, " *The Adelphi Papers* 38, no. 320 (1998): 1–89.

One theory indirectly linking poor economic development and political violence uses the rationale of economic opportunity.[32, 33] Insurgent organizations likely have an easier time recruiting among poor populations because they are able to provide financial incentives that benefit most individuals more than any other economic incentives available in a depressed area. That is, people living in very poor areas are likely to join an insurgency simply because it is the best financial decision available.[34]

Another widely accepted theory is that lower economic development is itself a proxy for a weak state. Countries that cannot provide good education and health care outcomes for their citizens are unlikely to be able to provide other crucial services, including security.[35] Weak states are less able to protect their governments and citizens against insurgent activity, providing more opportunities and incentives for insurgencies to grow and thrive. Finally, poor economic development and political violence often form a vicious circle whereby political violence, driven partly by poor development, further degrades development, ultimately leaving the country at a greater risk for future violence after one conflict ends.[36]

Finally, some theories of poor economic development and conflict processes focus on perceptions of economic inequalities among individuals and groups in society. Relative deprivation accounts for the former and horizontal inequality the latter. Ted Gurr's influential book, *Why Men Rebel*, proposes relative deprivation as the theory explaining how poor economic development contributes to conflict processes. The theory describes the mismatch between peoples' levels of expectation regarding their economic situations and the realities of their economic situations.[37] Occasionally, the chasm between a person's expectations and reality is quite large. People that have experienced sudden negative changes of fortune are particularly vulnerable to experiencing relative deprivation or feelings of anger or jealousy of peer groups who are more economically prosperous. Relative deprivation theory maintains that this sort of discontent can lead to organized rebellion. The major revolutions of France, Russia, and China, famously studied by Theda

Skocpol, fit this pattern. In each case, the state was failing to modernize as fast as its neighbors, leading to widespread dissatisfaction among elites who were in a position to observe their comparative failures.[38] However, like economic deprivation, relative deprivation is considered to be a general risk factor but only an indirect cause of political violence.

Relative deprivation describes the mismatch between peoples' levels of expectation regarding their economic situations and the realities of their economic situations. When relative deprivation occurs, individuals are more likely to participate in armed rebellion.

Ted Robert Gurr, *Why Men Rebel*, (London, UK and New York, NY: Routledge, 2011), 24.

More recently, scholars emphasized horizontal inequality as opposed to relative deprivation. Whereas relative deprivation concentrates on individual cognitive factors, horizontal inequality captures aggregate levels of inequality across the group.[39] Central to this theory is that humans, being inherently social beings, strongly identify with groups and evaluate their group's success relative to others. As a result, the theory ties together research linking ethnic political exclusion and poor economic development to conflict processes. The effect of horizontal inequality is more pronounced when ethnic groups have recently experienced a loss of state power. The presence of horizontal inequality dynamics in a country increases its risk for violent conflict.[40] Moreover, the violence is likely to be concentrated in the geographic areas where horizontal inequality is most prevalent.[41]

DEMOGRAPHICS

There is some evidence that certain demographic patterns are positively linked to political violence. One frequently cited pattern is proportionally large populations of young people in comparison with the adult population. The so-called youth bulge may have contributed to historical conflicts, including the European Revolutions of 1848, the rise of Nazism in Germany in the 1930s, and the American anti-war and civil rights protests in the 1960s.[42] Henrik Urdal examined a large historical dataset and found a relationship between nations with a comparatively large percentage of young people (fifteen to twenty-four years old) and levels of smaller scale political violence.[43] Moreover, the effects of the youth bulge are more pronounced under conditions of economic stagnation. When there are avenues for migration, however, the effects are less pronounced, suggesting that migration acts as a release valve for discontented youth.[44] However, another researcher notes that the impact of low income levels usually trumps the effect of the youth bulge in statistical models.[45]

Youth bulge is a demographic pattern in which a population has a disproportionately large youth population in comparison to the older population.

There are several theories about why a youth bulge is a risk factor for armed conflict. Unemployment and lack of economic opportunity are usually considered the most important elements. Societies cannot provide enough jobs for a sudden swell in young adults, so frustration and unemployment cultivate grievances against state authorities. Second, the presence of large numbers of military-age males (sixteen to thirty years old) provides a recruiting pool for existing resistance movements. A third factor may be that young people are more mobile and more likely to move to urban areas in search of employment, where urban overcrowding and the concentration of young, restive people may create conditions for street protests.[46] A fourth factor is psychological. Young people, especially those who are unattached, may be more likely than their elders to protest against grievances that are felt by all and may be more accepting of the risks associated with protest against the government, particularly autocratic regimes. Lastly, one unexpected factor comes from the expectation of youth revolts among state rulers. Because it is generally acknowledged that countries with a youth bulge are more prone to violence, rulers of young states are more likely to engage in repressive behaviors against them.[47]

The youth bulge theory is relevant to security concerns among regions prone to violence. In the first decade of the twenty-first century, about two-thirds of new civil war outbreaks occurred in demographically young countries. Currently, sub-Saharan Africa has a higher concentration of young people than any other region in the world, and its youth population growth rate is also the highest in the world. Altogether, nearly half of the world's population is under the age of thirty; of that global youth cohort, around 86 percent live in the developing world.[48]

REGIMES AND GOVERNANCE

A country's system of government, or regime type, is also a potential risk factor for armed violence. Regime type is typically scored on a spectrum from highly repressive regimes, sometimes called authoritarian or closed regimes, to fully democratic or open regimes. There is an abundant class of regimes residing in the middle of the spectrum, called anocracies, illiberal democracies, or hybrid regimes, that mix features of democracy and authoritarianism. Common sense might suggest that the most closed and repressive governments are the most at risk, but an analysis of the costs and benefits of violent and nonviolent resistance against open and closed governments presents a more nuanced picture.

It is generally agreed that consolidated democratic governments are the least vulnerable to violent opposition.[49] However, this does not mean that opposition is entirely absent in democratic governments. Open democracies offer nonviolent channels for opposition, including elections, public protests, and other forms of free speech that can be practiced within the legal framework of society. As a result, most opposition groups understand that there is little benefit in opposing the government through armed force when there are other less costly yet effective means of opposition available. Opposition groups interested in using violent tactics would also have difficulty gaining supporters for the same reason. Democracies are also thought to reduce violent conflict because democratic institutions offer greater avenues for bargaining with authorities and reducing commitment problems for reforms when groups oppose the state.[50]

By contrast, highly repressive regimes offer few legal means of resistance. The most repressive regimes prevent opposition groups from forming and are thus notably stable. These regimes usually have effective security forces that are free from legal restrictions on intelligence gathering, interrogation, or arbitrary arrest. Before the Arab Spring, the Middle East was home to some of the most repressive, but long-lasting, authoritarian regimes in the world. Scholars speculated that the primary reason Middle Eastern states held out so long against meaningful liberal reform was the robustness of the security forces, a highly effective, coercive apparatus resistant to political reform or transition to democracy.[51] In the case of Egypt during the Arab Spring, the refusal of the military to violently repress the masses of protesters in the country was the central feature facilitating President Hosni Mubarak's ouster.[52] In authoritarian countries with robust security forces, most dissent is quashed before it can gain momentum. For these reasons, highly repressive regimes, such as that in North Korea, are typically stable despite the grievances their citizens may hold against them.

Many modern states fall somewhere between consolidated democracy and the most repressive dictatorships. These blended regimes are called anocracies, hybrid regimes, or illiberal democracies. Blended regimes share features common to both authoritarian and democratic governments. For instance, an anocratic regime may allow opposition political parties to form and participate in elections but rig elections so that the ruling party is never seriously challenged. Anocracies are also described as states with weak central governments lacking effective policing and counterinsurgent components.[53] Many researchers interested in questions relating to the influence of regime type on civil war or other violent conflict often rely on the Polity Project dataset to measure a country's type of regime. The dataset assigns every country democracy ratings on a scale from most authoritarian to most democratic. Polity scores account for the following subcomponents: competitiveness of political participation, regulation of political participation, competitiveness of executive recruitment, openness of executive recruitment, and constraints on chief executive.[54]

Anocracies, hybrid regimes, or illiberal democracies share features common to both authoritarian and democratic governments.

The correlation between anocracies and civil war is the so-called inverted U-shaped curve because anocratic countries with middle-range Polity scores are more likely to experience civil war.[55] The instability of anocracies is attributed to the policies of hybrid governments that allow enough freedom of speech and assembly to allow opposition groups to form but are autocratic enough that challengers are often put down forcefully, causing opposition groups to believe they must resort to violence to achieve their aims. These hybrid governments:

> possess inherent contradictions . . . . [They] are partly open yet somewhat repressive, a combination that invites protest, rebellion, and other forms of civil violence. Repression leads to grievances that induce groups to take action, and openness allows for them to organize and engage in activities against the regime.[56]

Not only can regime type impact a country's risk for civil war, but it can also affect how the government rules its citizens. Governments that cannot fulfill the basic services typically attributed to the responsibility of state governments, such as the assurance of security and access to health care and education, are said to have poor governance capabilities. As previously discussed in the section on economic grievances, low levels of income or poor economic development are indicators that a country is more likely to experience a civil war than its more economically developed peers. Less certain, however, is why low income is associated with a higher risk for armed conflict. Using a theory based in economics, some speculate the effects occur through the labor market, meaning there are more individuals available for recruitment by armed groups and these individuals are more likely to positively respond to financial incentive.[57] The correlation between low income levels and civil conflict is probably not a direct one but is related to a third factor not included in the hypothesis. It might also be the case that low income indicates some other hidden factor associated with poor governance, including high levels of corruption, poorly performing security forces, poor rule of law, and low levels of accountability. This hypothesis is supported by data that show that countries rated poorly on governance tend to also have low income levels.[58] Several datasets are available for measuring the quality of governance, including the World Bank's Worldwide Governance Indicators and the Political Risk Services (PRS) International Country Risk Guide.[59]

GEOGRAPHY

Bad Neighborhoods and Conflict Traps

Countries with a history of violence are more likely to experience violence in the future.[35] The effect is so pronounced that most instances of civil war today are not first-time occurrences. About fifty years ago, most civil wars occurred in societies that had not experienced a previous conflict. Today, around 90 percent of civil wars are recurrences of previous conflicts.[60] The threat of recurring civil war is called the "conflict trap."[61] Even if a state itself has not experienced civil war, if its neighbor has, that state is at increased risk for conflict.[62]

Researchers in the social sciences have identified political and psychological reasons for the conflict trap. The simplest cause may be the available supply of weapons and people trained to use them, either in-country or nearby. When a conflict ends, weapons suppliers and soldiers remain, leaving enduring organizational legacies of violence that make future violent mobilization more likely.[63] Civil war violence can also exacerbate the economic and political conditions that contributed to conflict in the first place. Poorer, weaker states are more likely to experience violence than other countries.[64] Additional facets of conflict outcomes shape the potential for civil war recurrence, particularly how the previous conflict terminated. Civil wars that end through negotiated settlement, for instance, are more likely to recur than other types of termination, usually because both sides maintain the military capability to reignite the conflict.[65] Negotiated settlements are more likely to endure when they are secured by third parties and include power-sharing agreements, especially regarding security sector reform.[66]

The conflict trap refers to the tendency for countries that have experienced one civil war to break down into violent conflict after a period of peace.

Civil wars tend to cluster in certain regions, the so-called bad neighborhoods. A possible reason violence bleeds across borders is the large number of refugees or other displaced persons. These refugees may strain the resources of adjacent areas and be seen as economic competitors, leading to violence. The refugees may hold claims on their prior land (such as displaced Palestinians) or have other grievances (e.g., lost relatives and friends) to be redressed with violence in a new location.[67] Refugee flows can also alter the ethnic balance of host states, making it more likely that the conflict will spread there. This effect occurs when refugees have ethnic kin in their host state.[68]

However, psychological processes also contribute to clusters of regional violence. Most people have natural inhibitions toward the perpetration of violence, and most cultures have norms and practices designed to curb violence and its effects. In general, populations can become desensitized to violence over time, which begets more violence in the future. Children who witness adults behaving aggressively, for example, tend to imitate the aggression. Exposure to violence will lead some to commit more acts of violence, through desensitization or imitation, affecting how quickly people intervene or punish incidents and generally weakening cultural mores that prevent violence. Colombia, for instance, a country that has experienced decades of egregious insurgent violence, is described as a society in which violence is a "permanent characteristic," marked by high levels of homicides and assassinations even outside politically motivated violence.[69]

Countries that share borders with states experiencing a civil war are more likely to experience a civil war themselves. These regions are called bad neighborhoods.

Mountainous Terrain and Peripheries

The geographic features of a country are often cited as risk factors for conflict, whether it is slope elevation, mountainous terrain, or rural countryside. In the past, most resistance movements either began, or solidified, their presence in the countryside on the periphery of state power.[70, 71] Rough terrain is a typical topographical feature correlated with rebel activity as it provides safe havens and resources for insurgents.[72, 73] For instance, it features in Che Guevara's *foco* theory, in which rural-based guerrillas extend operations from the periphery of the country to its center of state power to overthrow the incumbent regime. The small bands of guerrillas provide a focus for a society's grievances against the regime, preparing the way for a mass-based insurgency or civil war. More recently, insurgent groups such as the Afghan Taliban benefited from the mountainous terrain of Afghanistan, making pursuit and surveillance by countervailing forces difficult. Likewise, the Vietcong in Vietnam benefited from dense forest cover despite American attempts at defoliation.[74] Several studies found a relationship between mountainous terrain and the incidence of civil war in a country,[75] but a consistent relationship between forests and insurgent activity has not been found.[76] Less clear are the reasons behind the relationship. Most theories explaining this relationship center on insurgent viability and a state's capacity to govern.

Colombia: Rugged, Mountainous Terrain and Sustained Insurgency

Colombia's physical environment is a central factor in the violence that continues to plague the country today. The country is infamous for its imposing terrain. At the northern end of the Andes mountain range, Colombia's landscape is dominated by commanding peaks—some reaching a height of seventeen thousand feet. The average peak in the country is a more modest nine thousand feet. Its rugged, mountainous interior shaped settlement patterns and troubling political legacies contributed to both historical and ongoing cycles of violence in the past and in the contemporary era. Its rugged landscapes offered safe havens for numerous insurgent and paramilitary groups. Most notably, Colombia's geography, alongside its political history, played a key role in the development of its weak central state. Colombia's weak state capacity presents a threefold danger to its political stability—it foments grievances in underserved areas, allows the emergence and sustainment of insurgent organizations, and gives rise to armed self-defense and paramilitary groups.

In the modern era, physical environments encompass more than a country's natural landscape. In the past century, many countries, including Colombia, have witnessed historically unprecedented rural migrations to urban environments. Most of the world's population now lives in cities, not the rural countryside. The FARC and the Ejército de Liberación Nacional (ELN), two of the most enduring leftist guerrilla groups in the country, developed strategies that relied heavily on exploiting the rural hinterlands and the populations those lands supported. The groups relied on the inaccessible geography for headquarters, training camps, and safe havens to evade Colombian security forces. Another guerrilla organization, the Movimiento 19 de Abril (M-19), followed the urbanizing trend instead, adopting an urban-based strategy. Operating in the city presented difficulties for the group, especially in regard to operational security, given the high state intelligence penetration within urban environments such as Bogotá, Cali, and Medellín. Whether in terms of organizational structure or military strategy, the physical environment in which the insurgents live impacts how the insurgents, and counterinsurgents, operate.

Katherine Raley Burnett, Christopher Cardona, Jesse Kirkpatrick, Sanaz Mirzaei, and Summer Newton, *Case Studies in Insurgency and Revolutionary Warfare: Colombia (1964–2009)* (Fort Bragg, NC: USASOC, 2012), 17–30.

The most frequently cited mechanisms connecting rough terrain and conflict are the positive effects rough terrain has on insurgent viability and the negative effects on state power. As in the cases previously mentioned, rough terrain can provide a safe haven for rebel groups, providing coverage from detection and aerial attack. Groups such as the FARC in Colombia and the Abu Sayyaf in the Philippines have used such forested areas as safe havens. Forests also provide valuable food and timber resources that can support insurgencies. The Karen National Union insurgents in Burma continue to exploit the country's valuable teak resources to help finance the world's longest-running insurgency.[77] The impact of forest cover on insurgent activity, while readily apparent in Burma, is confined to several states, rather than a general trend impacting conflict patterns across a great number of cases.

Rough terrain hinders a state's ability to effectively operate within its own territory. Areas of rough terrain may be far from the capital and lack the basic infrastructure, such as roads, necessary for the state to access and maneuver in these areas. As a result, as in some mountainous areas in Afghanistan, the government has little, if any, authority over residents there. This absence of the state can create opportunities for insurgencies to emerge and flourish. Poor state capacity in these regions can also contribute to socioeconomic conditions that fuel rebellion, primarily the lack of infrastructure, the lack of basic social services such as health care and education, and poor economic opportunities. Many states have rough terrain, and not all of these states are embroiled in conflict. Conversely, some regions that experience conflict, such as Anbar Province in Iraq, are flat and open. Also, some insurgencies, like the one in Northern Ireland, take place almost entirely in urban settings. One researcher describes the complex relationship thus:

> What the theory does predict is that rebels who seek refuge in the mountains are better able to withstand a militarily superior opposition . . . that rebel groups will take advantage of such terrain, whenever available. We do not believe that terrain in and of itself is a cause of conflict, nor does the rough terrain proposition anticipate such a relationship.[78]

In short, rough terrain is correlated with conflict, but that does not mean it causes conflict or that rough terrain is necessary for a conflict to emerge. In another study, researchers found that insurgent activity was more likely to coincide with areas of strategic importance than with areas that offered simple safety. Rough terrain, whether mountains or forest, is suitable for safe havens but might be too far from important targets to offer much strategic advantage for an attacking force. In a study of conflict in six African states, researchers found that states' so-called "pressure points"—areas "proximate to important strategic and economic sites," whether airports, areas with high road density, or large population centers—demonstrated the highest risk for experiencing insurgent activity, not rough terrains.[79]

Other geographic features, like location and distance, have also been found to impact conflict patterns and processes. Generally, research shows that regions farther from the capital are at higher risk for conflict, as well as those closer to international borders.[80] Proximity to international borders is a risk factor for conflict because insurgents often use neighboring countries, either willingly or unwillingly, as a safe haven from state forces. Moreover, conflicts between evenly matched belligerents are more likely to occur in proximity to the capital, whereas asymmetrical contests are usually confined to the peripheries far from the center of power.[81] The effects of state inaccessibility on conflict processes are more pronounced when married with eth-nopolitical exclusion of the affected communities.[82] The mechanisms underlying the relationship between geographic location and insurgent activity are not clear cut. Most often, the relationship is explained in terms of declining national strength as the government travels farther from its seat of power in the national capital. Kenneth Boulding called this the loss of strength gradient (LSG). As a result, the state has less capacity and presence in peripheral regions, providing an opportunity for rebels to emerge and grow.[83] The LSG captures the importance of poor military presence in peripheral regions but also a lack of state capacity to provide adequate social provisions, such as education and health care, that might contribute to the emergence of violence.[84]

The LSG predicts that states have less capacity to assert their power the farther away from the state capital.

POLITICAL OPPORTUNITIES

The theory of political opportunities is a quasi-structural condition that explains different components of the resistance process by focusing on the interactions between resistance actors and their environment. It also focuses on both violent and nonviolent resistance, although it was first used to explain the latter form of resistance. One of the advantages of this approach is that the focus is relational; the theory provides a good conceptual toolbox for analyzing a resistance as an unfolding process, rather than as a static phenomenon. The underlying theory of the political opportunity concept is that factors in the political environment incentivize individuals or groups to make decisions or take actions that they otherwise might not have had those opportunities not been present.[85]

> *The concept of political opportunity highlights factors in the political environment which incentivize individuals or groups to make decisions or take actions regarding participation in resistance that they otherwise might not have had those opportunities not been present.*

Sidney Tarrow, one of the pioneers of this theory, notes that history is replete with people and societies that have suffered from deprivation, oftentimes at the hands, directly or indirectly, of institutional authorities. The incidence of resistance against those authorities, such as the French Revolution, is quite rare. One of the differences, Tarrow argues, between those instances when resistance might have occurred but did not and those instances when it did occur, is political opportunity:

> What does vary widely from time to time and place to place are the levels and types of opportunities people experience, the constraints on their freedom of action, and the threats they perceive to their interests and values.[86]

However, while political opportunities may explain why groups mobilize at some junctures or adopt certain strategies, opportunities are a necessary, but not sufficient, condition for resistance groups to succeed against their opponents.[87]

The theory of political opportunity emerged at a time when most social science researchers viewed resistance as a primarily individual psychological process. Resistance was viewed as an irrational endeavor driven by changes or weaknesses in society that prompted psychological responses resulting in collective action. The psychological responses were alternatively characterized as resulting from structural strains that produced feelings of alienation, anxiety, or hostility. Once those feelings reached a certain threshold, organized resistance resulted. In this classical model of resistance, participants were not seen as aiming to achieve political goals so much as seeking a means to manage uncomfortable psychological conditions. These views gave way to more nuanced concepts of resistance that viewed it as a rational endeavor when it became clear, particularly during the 1960s, that resistance could produce meaningful and lasting changes in government policies and social values. The emphasis also shifted from individuals in resistance to the organizations individuals formed to achieve their objectives, as well as motivations outside structural strain.[88]

In response to criticisms of the theories of resistance previously described, Doug McAdam proposed a new model for explaining resistance, called the political process model, which helped propel research on political opportunity. The model described resistance as the culmination of long-term political processes dictated by the existing power configurations in society, rather than the dramatic, short-term processes depicted in the classical strain model. He identified three key factors in the political process model that explained the rise and decline of resistance movements, including the level of organization among the relevant population, positive assessment for the success of the insurgency, and the configuration of political actors within the government.[89]

The political process model describes resistance as the culmination of long-term political processes dictated by the existing power configurations in society.

Political Opportunity: US Civil Rights Movement

Doug McAdam applied his political process model to black protest movements in the United States in the period surrounding the civil rights movement. He used the model to explain the movement's rise and eventual decline. He found that there were several political opportunities that facilitated the rise of the African American protest movement: the Supreme Court ruling on *Brown versus Topeka Board of Education, the decline of the cotton economy, and a drop in the number of black lynchings. The shifts in political opportunity increased a sense of efficacy among African American populations and encouraged greater organizational efforts. McAdam described the emerging sense that changing the political and social status quo was as possible as cognitive liberation. The efforts contributed to the growth of social networks in the black churches, black colleges, and southern chapters of the National Association for the Advancement of Colored People that were integral to the success of the civil rights movement.*

Doug McAdam, *Political Process and the Development of Black Insurgency, 1930–1970* (Chicago, IL and London, UK: University of Chicago Press, 1982), 40–42.

Eventually, the concept of political opportunity evolved to incorporate three broad categories of variables, called the political opportunity set. The first, political opportunity structures, refers to the formal or permanent dimensions of the environment that shapes incentives for resistance. The second, the configuration of actors, assesses the existing relationships between powerful actors in the environment. The actors include a resistance group's potential allies, its adversaries, and influential bystanders. Lastly, the political opportunity set includes the dynamic process of ongoing interaction between resistance groups and their adversaries that impact the group's strategies, tactics, and political objectives or claims. Each of the variables are powerful tools for better understanding resistance processes and their outcomes.[90]

The political opportunity set is comprised of three interrelated concepts that explain the political context surrounding conflict processes. The concepts include political opportunity structures, the configuration of actors, and dynamic interaction.

The political opportunity structures capturing the formal or permanent dimensions of the environmental context can include a regime's openness, its capacity for carrying out its dictates, and the methods with which it handled challengers in the past. Some of these factors have already been discussed, where the levels of democratization or authoritarianism in a regime impact how a resistance movement develops according to the costs and benefits associated with each end of the spectrum. However, the contentious politics framework teases out the distinctions driving the authoritarian and democratic responses to resistance and incorporates studies of violent and nonviolent movements. Charles Tilly and Sidney Tarrow, for example, use two measures to identify critical characteristics of a regime. The first, a regime's level of democracy, measures how open a political system is to challengers. Established democracies tolerate a great deal of challenge so long as the resistance follows the rules of the game (i.e., no violence). The second, a regime's capacity, refers to the ability of a state to enforce its policies. A state may wish to be more authoritarian and crush challenges to its rule, but it may lack the coercive capacity to carry out its preferred response.[91]

In general, regimes that have high levels of democracy combined with high capacities are least likely to experience organized violence but probably experience a great deal of civil resistance through more-or-less legitimate channels. By contrast, regimes with low levels of capacity and low levels of democracy are in a zone called fragmented tyranny, where violent resistance is more likely to occur than other types of resistance. It is in these regimes that most civil wars take place. Likewise, resistance in high-capacity, undemocratic regimes is more likely to be clandestine and characterized by brief confrontations met with repressive measures by security forces.[92] When some of these regime characteristics shift in some way, the opportunities available to resistance actors also shift, which can explain why some resistance movements opt to mobilize or adopt certain strategies and tactics at a particular point in time. States undergoing the process of democratization or de-democratization, for instance, signal the presence of shifting alignments among elites and stakeholders that resistance actors can leverage to gain powerful allies that support them.[93]

The political opportunities available to resistance movements are also shaped by the configuration of three important sets of actors in a political environment: allies, adversaries, and bystanders. There are many potential allies available to resistance movements that can act as a force multiplier. Some allies might be sympathizers, important policy makers in a targeted government, material or ideological support by foreign governments, or powerful interest groups able to apply pressure on government authorities to meet resistance demands. Increasingly, well-funded and well-respected nongovernmental organizations can also be influential allies. Adversarial actors also cover a wide range, some more unexpected than others. Most often the primary adversaries are government authorities or security forces. However, adversaries can also be countermovements that are not necessarily part of the incumbent administration. The longrunning leftist insurgency in Colombia, for instance, has for a long time battled paramilitaries that support the regime. The paramilitaries, with a fluctuating level of collaboration with the Colombian government, have acted as a spoiler in peace processes by assassinating leftist insurgents demobilizing into the legitimate political system.[94] In addition, adversaries might also be other resistance groups that, despite having similar objectives, are fierce competitors for limited resources and popular support. Lastly, the bystanders are the observers of unfolding events in the domestic and international setting. Understanding the configuration of these key actors provides insight into the incentives, or lack of them, for mobilization.[95]

Mapping the configuration of actors requires analyzing the actors' capabilities, interests, and likely payoffs for particular courses of action. During the Arab Spring, the militaries of Egypt, Syria, Libya, and Tunisia made different decisions regarding the repression of protesters. The decision made by Libyan and Syrian military leaders to (mostly) not defect was driven in large part by the expected payoffs of regime survival. In both countries, the military was closely intertwined with the Assad and Qadhafi regimes. Both rulers tightly integrated the military into the political regime by delegating power to extended kin and social networks. The survival of the military leaders in each country was closely tied to the fate of the ruling regime. Egyptian President Hosni Mubarak, by contrast, secured the military's loyalty through the distribution of extensive economic privileges; the military elite were among the wealthiest members of Egyptian society. However, in the years before the Arab Spring, Mubarak funded his own internal security units at the army's expense when the threats to his rule shifted from external to domestic ones. Moreover, military leaders deeply resented Mubarak's plans to transition his rule to his son, Gamal Mubarak. With the looming threat of losses in economic and political privileges, the military broke with its tradition of support to President Mubarak. As a result, when the protesters took to the streets in Tahrir Square, the military refused Mubarak's command to fire on protesters.[96, 97]

By contrast, the Syrian military had ample incentive to protect Assad's rule, making it more likely that the military would actively fire on protesters when Assad issued the order. The Egyptian military, like most of the Egyptian population, is Sunni. In the Syrian military, on the other hand, the minority Alawites, President Bashar Assad's communal group, are heavily represented, sometimes comprising about 90 percent of higher officers, while Alawis as a group comprised only about 10 to 12 percent of the Syrian population.[98] Moreover, the Assad regime also used the distribution of material resources through patronage as a cornerstone of his coup-proofing policy. He primarily favored officers, prompting many charges of corruption among the enlisted.[99]

Despite the divisions Assad's patronage caused among the armed forces, the military did not defect in great numbers. The military's continued support of the Assad regime stems from its communal connections and the economic incentives afforded even Sunni officers. It had little incentive to expect its power and influence to continue in a Sunni-dominated political regime or one that did not continue to provide desired economic incentives.[100] An examination of the configurations of actors in this case sheds light on the incentives Arab militaries did or did not have to protect the incumbent regime. The decision was crucial for the success of nonviolent resistance in Egypt and the emergence of violent resistance in Syria after the military's armed repression of protests.

Shifts in actor configurations are especially critical to the success of nonviolent resistance campaigns. One of the most effective tools of resistance actors is to help generate dissensus among the political elite to gain powerful allies, called an elite fracture. The mass participation that usually accompanies successful cases of nonviolent resistance is an optimal tool for leveraging elite defections among political and economic stakeholders in the adversarial regime. In nonviolent resistance campaigns, one of the most important pathways to success is encouraging security forces to refuse governmental orders demanding violent repression to neutralize resistance movements' challenges to state authorities.[101] Military defections, or nondefections, are among the most crucial reasons some groups' nonviolent campaigns fail and others succeed.[102]

The 1979 Iranian Revolution: Configuration of Actors and Military Defections

In 1979, Ayatollah Khomeini established a theocratic regime in Iran after unseating Shah Reza Pahlavi, whose family had ruled as Iranian monarchs since 1925. Khomeini's rise to power resulted in large part from a nonviolent resistance campaign that garnered mass participation. He was the head of a coalition of secular politicians and Islamic clerics backed by other influential groups in society. Khomeini helped organize massive protests against the Shah in major cities like Tehran and Tabriz. The shah's opponents were unhappy with his political, economic, and religious policies that sought to integrate Iran more closely with the capitalist, secular West. While some of his policies initially stimulated development in the country, several deep economic recessions frustrated the growing middle class. After a series of massive protests and strikes, Khomeini succeeded in driving the shah from power.

The shah invested heavily in the Iranian military, which he regarded as insurance against internal and external threats. Although security forces used force on several occasions, the military officers made the crucial decision not to actively repress the shah's challengers. Opposition leaders were aware that the decisions of military leaders were critical to the success of the resistance movement. They regularly met with police and military officials, entreating them to join the opposition or at least not to fire on protesters at the shah's command. During protests, participants favorably engaged the shah's military, giving soldiers flowers and offering civilian clothes to those who wished to join the protest. The opposition leaders were effective in driving a wedge between the political and military elite, which was measureable in the high levels of desertions and leave requests, reportedly lowering morale in the ranks of police and the army. Loyal officers attempted to isolate the security forces from protesters, with little effect. Outmaneuvered by the regime's opponents, the shah's bloated military forces did not leave the barracks to secure the leader's position.

Summer Newton, "1979 Iranian Revolution," in *Casebook on Insurgency and Revolutionary Warfare, Volume II: 1962–2009* ed. Chuck Crossett (Fort Bragg, NC: USASOC, 2012), 78–104; Erica Chenoweth and Maria J. Stephan, *Why Civil Resistance Works: The Strategic Logic of Nonviolent Conflict (New York, NY: Columbia University Press, 2011), 106–108.

INTERNATIONAL DIMENSIONS

Many theories of political violence and civil war have seen such events as being primarily domestic affairs, shaped by the particular grievances, opportunities, and regime structures of the affected country. However, a growing body of evidence strongly suggests that resistance is not only shaped by domestic variable but is also affected by the international and regional environment. Foreign governments, both in the region and farther abroad, often take sides in civil conflicts and provide critical resources that resistance organizations need to survive. In addition, insurgent groups may find sanctuaries in neighboring states and take advantage of porous borders, providing them safe areas in which to operate. Conflicts can also spread to neighboring countries, particularly as refugee flows, ethnic kinship ties, and transnational militant groups destabilize entire regions. These international dimensions have important implications for the onset, conduct, and duration of the conflict.

The political opportunity framework, previously discussed, stresses the importance of allies in shaping the trajectory of resistance movements. One important potential ally group is that of foreign governments, which have unique capabilities to provide financial resources and military equipment to insurgent groups. Target governments that have an international rival are far more likely to be involved in a civil conflict, as rivals often seek to indirectly destabilize the state through rebel proxies.[103] Such foreign ties have pros and cons for the sponsoring state and for the resistance group. One useful perspective for understanding such relationships is principal-agent theory. The principal, or foreign sponsor, provides resources and support in exchange for some control over the agent's agenda. The sponsor may try to shape the ideology of the group, the types of targets it selects, and its negotiation strategies with the government. The foreign government gains a relatively low-cost way to destabilize and place pressure on its rival, as it does not risk its own military forces. However, it can also lose control over the agent if it takes actions that are unwelcome from the perspective of the sponsor; in extreme cases, the agent can turn on the principal. For the militant group, it gains vital resources quickly but can lose decision-making autonomy if it becomes overly dependent on foreign support. It may also lose domestic legitimacy if it comes to be seen as a pawn of an external power.[104]

Such relationships were quite prevalent during the Cold War, as the United States and the Soviet Union competed for influence in the developing world. The Soviet Union and its allies backed insurgencies in places such as Cuba, Rhodesia (Zimbabwe), Vietnam, and Central America. The United States supported anti-communist forces in Afghanistan, Nicaragua, and Angola, among others. At times, these groups engaged in unwelcome activities, such as targeting civilians, and some elements within these resistance movements became hostile to US interests. Other governments have been responsible for supporting insurgencies as well. Iran provides material backing for Shia resistance organizations such as Hizbollah in Lebanon; Rwanda supported various militant groups in the Democratic Republic of the Congo; and Pakistan assists anti-Indian groups. Sometimes such a strategy is quite successful from the perspective of the sponsor, but principal-agent theory also stresses the potential for agency loss, or the distrastrous consequences of losing control over the agent. For example, while Rwanda successfully toppled the government of Mobutu Sese Seko in Zaire (the Democratic Republic of the Congo) by supporting the forces of Laurent Kabila, the agent ultimately turned on the principal, sparking a major international conflict that drew in multiple states in Africa.[105]

One resource that neighboring states in particular can provide is sanctuary across an international border.[106] At times, external bases are willingly provided by the host state, as when Jordan allowed the Palestine Liberation Organization (PLO) access to its territory. Other times, rebels can take advantage of weak states and porous borders, without the complicity of the host, such as the Islamic State of Iraq and Syria (ISIS)' ability to move freely between Iraq and Syria. Previously, the importance of mountains, difficult terrain, and peripheral areas were discussed as important geographical features that help facilitate resistance organizations. Sanctuary in neighboring states provides additional cover as counterinsurgents cannot easily pursue groups across an international border without offending their neighbor and have relatively poor intelligence gathering capabilities in foreign states. Such sanctuaries help violent movements survive longer than they would otherwise, but they can also spark international conflicts between host and target governments.[107]

Scholars have also noted that violence can spread quickly across a region as instability in one state may spark instability elsewhere, a process known as diffusion.[108] Three particular mechanisms help explain this phenomenon. First, resistance movements in one country can learn by example from successful movements elsewhere, particularly those in the same regional and cultural context.[109] This can embolden militants and activists and teach them strategies of effective resistance. The Arab Spring, for example, was inspired by successful protests in Tunisia and Egypt, even if movements elsewhere could not repeat the same success.

Second, ethnic and religious groups often span national boundaries, and kinship ties can extend networks of resistance. A shared experience, cultural and familial ties, the cross-border exchange of resources and information, and even common organizational structures can facilitate the spread of resistance to neighboring states.[110] For example, ethnic Albanians in Kosovo fought for independence against Serbia in the late 1990s, and the Kosovo Liberation Army shared weapons, information, and even key personnel with Albanian militants in nearby Macedonia. In addition, some Islamist movements such as ISIS and al-Qaeda, have numerous affiliates among coreligionists across multiple states, uniting jihadist organizations under a common umbrella.

Finally, refugee flows have been argued to help facilitate the spread of conflict to nearby states, particularly if host governments have poor institutional capacity to absorb large numbers of migrants.[111] Refugees

51

can place a financial burden on host governments, be viewed as economic competitors, and shift the demographic/cultural balance in host communities. Refugees in squalid camp conditions, with few avenues for meaningful livelihoods, may find participation in militant activities to give them better economic opportunites and a sense of purpose. Finally, militants often attempt to capture humanitarian resources intended for refugees, such as food and medical supplies, as aid workers and nongovernment officials are easy targets.[112] Therefore, rebels often use refugee camps as sources of recruits and supplies, facilitating conflict in the target state, but also threatening to escalate tensions in the host state. As such, it is vital that the host state and the international community take steps to provide security to refugees, prevent militant access to camps, and assist in integrating refugees among host communities.

While these international dimensions may help spark new conflicts, they also have important implications for conflict resolution, which is covered in greater detail in chapter 5. The introduction of new actors, such as foreign governments and ethnic kin groups, can add a layer of complexity to attempts to negotiate a settlement.[113] Not only must the primary belligerents agree to a settlement, but regional actors can act as spoilers, or work to block a deal, if their preferences are not satisfied as well. Attempts to bring actors to the bargaining table in Syria, for example, have been hampered by external parties both on the government and on the insurgent side, who are pursuing their own interests. This has the effect of making conflicts with significant external involvement significantly longer and more resistant to conflict-management strategies.

CONCLUSION

As the discussion in this chapter demonstrated, political violence is a highly complex, dynamic process that defies a singular explanation. There is no single theory that can explain why political violence or resistance occurs in some places but not others. Although many people would like to be able to predict when and where such violence occurs, the complexity of human interaction prevents such robust predictive power. Nevertheless, social scientists have spent decades researching which factors increase the probability of terrorism, insurgency, or other forms of political violence appearing in a state or region.

While there is no universal law or theory capable of predicting when and where political violence or resistance will emerge, the available research points to factors that make some states more susceptible to political violence than others. One promising area of research that moves beyond the impact of macrostructural factors is the contentious politics program. This research program more deftly captures the interactions of actors involved in the conflict to enable a better understanding of how the unfolding dynamics impact conflict processes. It is arguably of particular interest to the Special Forces because it includes the interplay between larger structural conditions and individual and organizational actions.

The broad factors associated with increased probability of conflict are called macrostructural variables. A great deal of the current research on political violence in the social sciences addresses the investigation of these factors. They capture the macro, or large-scale, institutions, conditions, or patterns in a state that impact how organizations and individuals act. Insurgencies and resistance movements form and operate in areas shaped by geographic, social, economic, and political systems. These features impact the decisions, behaviors, and actions that influence a group's formation, strategies and tactics, longevity, and chances for success.

One set of macrostructural variables captures how societal grievances can motivate organized resistance or political violence. The socioeconomic variables discussed here include ethnic and economic grievances, although the two sometimes overlap. Ethnic grievances are generated by the economic or political exclusion of ethnic groups. Although early research pointed to ethnic diversity as the culprit, it is more likely that some ethnic configurations are more prone to produce violence, such as when a majority group wields political and economic power over other marginalized groups on the periphery. Ethnic identity itself is fluid, often changing as a result of ongoing conflict processes. Economic grievances function similarly. Poor countries are more likely to experience conflict than more prosperous countries. It is not clear why poverty has such a powerful impact on political stability, but it is likely related to state weakness or the availability of a greater recruiting pool for insurgent groups. Poor states have more difficulty rebuffing armed opposition, and unemployed, poor individuals are more likely to be attracted by financial incentives offered by insurgents.

The onset of political violence is also attributed to a state's poor governance. Certain types of political regimes are more likely to generate conflict than others. Consolidated democracies, such as those common in the West, are the least likely to experience conflict. However, countries residing somewhere between closed authoritarian and democratic, open regimes are more likely to see political violence. These anocracies have some features of democracies, such as legal political opposition parties, but combine those features with authoritarian tendencies to repress state challengers. The quality of governance, not just the type of regime, is also a factor to consider. A government's inability to fulfill the basic functions of a state, such as ensuring citizens' security or offering health care, also contributes to the likelihood of violence. Poor governance feeds into factors already discussed, including levels of economic development and state weakness.

A country's geography and neighborhood also impact the probability of violence. States that have experienced a prior civil war, or have a close neighbor that has, are more likely to experience civil war in the future. The contribution of a violent past, and a violent neighborhood, is referred to as the conflict trap. Prior civil wars devastate the local economy and leave organizational legacies of violence that include weapons, fighters, and a political culture inured to violence. Countries that have mountainous, rugged terrain are also primed for insurgent violence. It is thought that the remote areas inaccessible to state security forces offer a safe haven for insurgencies to emerge and sustain themselves.

The political opportunity theory, part of the contentious politics program, has propelled a more fine-grained analysis of individual and organizational action. Research on macrostructural factors generally fails to capture the motivations of individuals and organizations involved in conflict processes. Many states might be poor, have mountainous terrain, or politically exclude some ethnic groups, but not all such states experience civil war or organized political resistance. Political opportunity theory seeks to address this gap by looking at political opportunity structures, the configuration of actors, and the ongoing interaction between resistance actors and their opponents. In this manner, political opportunity theory captures the interplay between broad macrostructural factors and individual action. Political opportunity structures can include existing or shifting qualities of a regime, such as changes in government policy that allow for political opposition to legally form. The configuration of actors is especially attuned to the relationships among diverse elite political actors, such as the ruling administration and its military. Gaps or shifts in these relationships can be powerful leveraging tools for resistance actors. Moreover, interactions between resistance actors and their opponents, generally the state, impact how the conflict processes unfold. Repeated interactions of state repression can lead a nonviolent resistance organization to adopt more violent tactics and strategies.

While a great deal of research has investigated the domestic factors influencing political violence, including grievances, opportunities, and regime structures, the international and regional environment also shape the onset, severity, and duration of civil wars. Foreign governments, whether located in the region of the conflict or farther afield, often provide external support for one or more actors involved in civil conflicts. With significant resources and assets at their disposal, states are in a position to supply resistance movements with the resources they needed for their survival. In addition, resistance movements also find witting or unwitting sanctuaries in nearby states with porous borders, offering them safe areas in which to plan and operate. Finally, the diffusion of conflict from one state to another is driven by refugee flows, ethnic kinship ties, and transnational militant groups that can destabilize entire regions, making lasting peace elusive.

ENDNOTES

[1] This chapter is adapted, revised, and updated from Nathan Bos, editor, *Human Factors Considerations of Undergrounds in Insurgencies*, 2nd ed. (Fort Bragg, NC: USASOC, 2013).

[2] Adria Lawrence and Erica Chenoweth, "Introduction," in *Rethinking Violence: States and Non-State Actors in Conflict*, ed. Erica Chenoweth and Adria Lawrence (Cambridge, MA: The MIT Press, 2010), 5–6.

[3] David S. Meyer, "Protest and Political Opportunities," *Annual Review of Sociology* 30 (2004): 125–145.

[4] The ARIS project involves research on both violent and nonviolent resistance as discussed in chapter 1, which details the ARIS conceptualization of resistance. However, the research related to macrostructural variables almost always specifically relates to violent resistance in the form of full-blown civil wars or low-intensity armed conflict. Over time, the ARIS project aims to expand on research that compares the differences between the onset of violent and nonviolent conflict.

[5] A time series analysis includes a sequence of data points over a continuous time interval. Therefore, a time series might cover the incidence of civil war in the period 1945–2015. A

cross-national series includes data across multiple countries. Many datasets, for instance, cover all the countries in the world with a population above a certain number.

[6] The UCDP/PRIO ACD dataset is available at http://www.pcr.uu.se/research/ucdp/datasets/ucdp_prio_armed_conflict_dataset/. See James D. Fearon, "Governance and Civil War Onset," *World Development Report 2011 Background Paper*, (31 August 2010): 6–9 https://openknowledge.worldbank.org/bitstream/handle/10986/9123/WDR2011_0002.pdf.

[7] Some articles compare results from different datasets and measures in the literature. See Adrian Florea, "Where Do We Go from Here? Conceptual, Theoretical, and Methodological Gaps in the Large-N Civil War Research Program," *International Studies Review* 14, no. 1 (2012): 78–98; Håvard Hegre and Nicholas Sambanis, "Sensitivity Analysis of Empirical Results on Civil War Onset," *Journal of Conflict Resolution* 50, no. 4 (2006): 508–535; Fearon, "Governance and Civil War Onset," 6–9, https://openknowledge.worldbank.org/bitstream/handle/10986/9123/WDR2011_0002.pdf.

[8] Elaine K. Denny and Barbara F. Walter, "Ethnicity and Civil War," *Journal of Peace Research* 51, no. 2 (2014): 199–212.

[9] Nils B Weidmann, Jan Ketil Rød, and Lars-Erik Cederman, "Representing Ethnic Groups in Space: A New Dataset," *Journal of Peace Research* 47 no. 4 (2010): 491–499.

[10] The ELF index relies largely on data from the Atlas Naradov Mira, a reference created by Soviet ethnographers in 1964. The Atlas is a comprehensive chart of ethnic groups worldwide.

[11] Philip G. Roeder, "Ethnolinguistic Fractionalization (ELF) Indices, 1961 and 1985," accessed March 1, 2016, http://pages.ucsd.edu/~proeder/elf.htm.

[12] Lars-Erik Cederman and Luc Girardin, "Beyond Fractionalization: Mapping Ethnicity onto Nationalist Insurgencies," *American Political Science Review* 101, no. 1 (2007): 173–185; Tanja Ellingsen, "Colorful Community or Ethnic Witches' Brew? Multiethnicity and Domestic Conflict During and After the Cold War," *Journal of Conflict Resolution* 44, no. 2 (2000): 228–249.

[13] Donald L. Horowitz, *Ethnic Groups in Conflict* (Berkeley, CA: University of California Press, 1985), 55.

[14] Frederik Barth, "Introduction," in *Ethnic Groups and Boundaries*, ed. Frederik Barth (Long Grove, IL: Waveland Press, 1998), 15–16.

[15] In some cases, Brubaker notes that the insistence of academics on coding conflicts as "ethnic" can serve as legitimation for leaders involved in the conflict.

[16] Rogers Brubaker, *Ethnicity Without Groups* (Cambridge, MA: Harvard University Press, 2004), 17–18.

[17] There are theoretical distinctions between nation and state, but for the sake of brevity, those distinctions are not introduced. In alignment with current common usage, the terms nation and state, unless otherwise specified, are used interchangeably in this text.

[18] Ernest Gellner, *Nations and Nationalism* (Ithaca, NY and London, UK: Cornell University Press, 2008), 1.

[19] Andreas Wimmer, Lars-Erik Cederman, and Brian Min, "Ethnic Politics and Armed Conflict: A Configurational Analysis of a New Global Data Set," *American Sociological Review* 74, no. 2 (2009): 316–337.

[20] Lars-Erik Cederman, Andreas Wimmer, and Brian Min, "Why do Ethnic Groups Rebel? New Data and Analysis," *World Politics* 62, no. 1 (2010): 87–119; Lindsay Heger and Idean

Salehyan, "Ruthless Rulers: Coalition Size and the Severity of Civil Conflict," *International Studies Quarterly* 51, no. 2 (2007): 385–403.

21 Other researchers have also combined geocoded inequality and ethnic settlement data. The measures of inequality include the use of satellite imagery of nightlight emissions which are highly correlated with economic activities.

22 Frances Stewart, *Horizontal Inequalities and Conflict*, ed. Frances Stewart, (New York, NY: Palgrave Macmillan, 2008). See also Lars-Erik Cederman, Nils B. Weidmann, and Nils-Christian Bormann, "Triangulating Horizontal Inequality: Toward Improved Conflict Analysis," *Journal of Peace Research* 52, no. 6 (2015): 806–821.

23 Relative deprivation describes the mismatch between peoples' levels of expectation regarding their economic situations and the realities of their economic situations. When relative deprivation occurs, individuals are more likely to participate in resistance.

24 Lars Erik Cederman, Kristian Skrede Gleditsch, and Halvard Buhaug, *Inequality, Grievances, and Civil War* (New York, NY: Cambridge University Press, 2013), 31–32.

25 Geocoded data are tied to specific geographic locations using some type of geographic information system (GIS) software.

26 Paul Collier and Anke Hoeffler, "Greed and Grievance in Civil War," *Oxford Economic Papers* 56, no. 4 (2004): 563–595; James Fearon and David Laitin, "Ethnicity, Insurgency, and Civil War," *American Political Science Review* 97, no. 1 (2003): 75–90.

27 The term proxy is used to describe an indirect measure of a variable used in research. In this example, researchers are interested in using variables that are good indicators of state weakness. One way to indirectly measure a country's level of economic development is by measuring the health of its citizens, which in turn is impacted by a state's policies related to health care and its health care infrastructure, among others. In general, citizens in wealthier states have better health outcomes than those in poorer states. As a result, some research investigating how economic development impacts political stability relies on the proxy measure of infant mortality, which measures the likelihood that a child will die before he or she reaches one year of age.

28 Håvard Hegre, Tanja Ellingsen, Scott Gates, and Nils Petter Gleditsch, "Toward a Democratic Civil Peace? Democracy, Political Change, and Civil War, 1816–1992," *American Political Science Review* 95, no. 1 (March 2001): 33.

29 Lars-Erik Cederman and Luc Girardin, "Beyond Fractionalization: Mapping Ethnicity Onto Nationalist Insurgencies," *American Political Science Review* 101, no. 1 (February 2007): 173.

30 Jack A. Goldstone, Robert H. Bates, David L. Epstein, Ted Robert Gurr, Michael B. Lustik, Monty G. Marshall, Jay Ulfelder, and Mark Woodward, "A Global Model for Forecasting Political Instability," *American Journal of Political Science* 54, no. 1 (January 2010): 190–208.

31 Alex Braithwaite, "Does Poverty Cause Conflict?: Isolating the Causal Origins of the Conflict Trap," *Conflict Management and Peace Science* 33, no. 1 (2016): 45–66.

32 Recent research has pointed to links between social unrest, usually in the form of riots and demonstrations, and increases in food prices. However, others have demonstrated the political and social context are better at explaining political violence than agricultural output or commodity prices.

33 See Joe Weinberg and Ryan Bakker, "Let Them Eat Cake: Food Prices, Domestic Policy, and Social Unrest," *Conflict Management and Peace Science* 32, no. 3 (2015): 309–326; See also, Halvard Buhaug, Tor A. Benaminsen, Espen Sjaastad, and Ole Magnus Theisen, "Climate Variability, Food Production Shocks, and Violent Conflict in Sub-Saharan

Africa," *Environmental Research Letters* 10, no. 12 (22 December 2015): 125015; David Keen, "The Economic Functions of Violence In Civil Wars," *The Adelphi Papers* 38, no. 320 (1998): 1–89.

[34] Paul Collier and Anke Hoeffler, "Greed and Grievance in Civil War," *Oxford Economic Papers* 56, no. 4 (2004): 563–595.

[35] James Fearon and David Laitin, "Ethnicity, Insurgency, and Civil War," *American Political Science Review* 97, no. 1 (2003): 75–90; Edward Miguel, Shanker Satyanath, and Sergenti, "Economic Shocks and Civil Conflict: An Instrumental Variables Approach," *Journal of Political Economy* 112, no. 4 (2004): 725–753.

[36] Alex Braithwaite, "Does Poverty Cause Conflict?: Isolating the Causal Origins of the Conflict Trap," *Conflict Management and Peace Science* 33, no. 1 (2016): 45–66.

[37] Ted Robert Gurr, *Why Men Rebel*, (London, UK and New York, NY: Routledge, 2011), 24.

[38] Theda Skocpol, *States and Social Revolution: A Comparative Analysis of France, Russia, and China* (Cambridge, UK: Cambridge University Press, 2015).

[39] Gudrun Ostby, "Polarization, Horizontal Inequalities and Violent Civil Conflict," *Journal of Peace Research* 45, no. 2 (2008): 143–162.

[40] Lars-Erik Cederman, Kristian Skrede Gleditsch, and Halvard Buhaug, Inequality, Grievances, and Civil War (New York, NY: Cambridge University Press, 2013).

[41] Gudrun Ostby, Henrik Urdal, Mohammad Zulfan Tadjoeddin, S Mansoob Murshed, and Havard Strand, "Population Pressure, "Horizontal Inequality and Political Violence: A Disaggregated Study of Indonesian Provinces, 1990–2003," *The Journal of Development Studies* 47, no. 3 (2011): 377–398; S. Mansoob Murshed and Scott Gates, "Spatial-Horizontal Inequality and the Maoist Insurgency in Nepal," *Review of Development Economics* 9, no. 1 (2005): 121–134.

[42] Jack A. Goldstone, Revolution and Rebellion in the Early Modern World (Berkeley and Los Angeles: University of California Press, 1991), 24–27; Herbert Moller, "Youth as a Force in the Modern World," *Comparative Studies in Society and History* 10, no. 03 (1968): 237–260.

[43] Henrik Urdal, "A Clash of Generations? Youth Bulges and Political Violence," *International Studies Quarterly* 50, no. 3 (2006): 607–629; Nils Petter Gleditsch, Peter Wallensteen, Mikael Eriksson, Margareta Sollenberg, and Håvard Strand, "Armed Conflict 1946-2001: A New Dataset," *Journal of Peace Research* 39, no. 5 (2002): 615–637.

[44] Henrik Urdal, "The Devil in the Demographics: The Effect of Youth Bulges on Domestic Armed Conflict, 1950–2000," *Social Development Papers: Conflict and Reconstruction Paper* 14 (2004).

[45] James D. Fearon, "Governance and Civil War Onset," *World Development Report 2011 Background Paper*, 14–16, https://openknowledge.worldbank.org/bitstream/handle/10986/9123/WDR2011_0002.pdf.

[46] Jack A. Goldstone, "Population and Security: How Demographic Change can Lead to Violent Conflict," *Journal of International Affairs* 56, no. 1 (2002): 3–21.

[47] Ranghild Nordas and Christian Davenport, "Fight the Youth: Youth Bulges and State Repression," *American Journal of Political Science* 57, no. 4 (2013): 926–940.

[48] Elizabeth Leahy Madsen, Beatrice Daumerie, and Karen Hardee, "The Effects of Age Structure on Development: Policy and Issue Brief," *Population Alliance Initiative*, http://pai.org/wp-content/uploads/2012/01/SOTC_PIB.pdf.

[49] Harvard Hegre, "Democracy and Armed Conflict," *Journal of Peace Research* 51, no. 2 (2014): 159–72.

[50] James Fearon, "Why Do Some Civil Wars Last So Much Longer than Others?," *Journal of Peace Research* 41, no. 3(2004): 275–301.

[51] Eva Bellin, "The Robustness of Authoritarianism in the Middle East: Exceptionalism in Comparative Perspective," *Comparative Politics* 36, no. 2 (2004): 139–157.

[52] Eva Bellin, "Reconsidering the Robustness of Authoritarianism in the Middle East: Lessons from the Arab Spring," *Comparative Politics* 44, no. 2 (2012): 127–149.

[53] James Raymond Vreeland, "The Effect of Political Regime on Civil War Unpacking Anocracy," *Journal of Conflict Resolution* 52, no. 3 (2008): 401–425.

[54] The Polity dataset is available at http://www.systemicpeace.org/polityproject.html.

[55] Edward N. Muller, "Income Inequality, Regime Repressiveness, and Political Violence," *American Sociological Review* 50 (1985): 47–61; Patrick M. Regan and Sam R. Bell, "Changing Lanes or Stuck in the Middle: Why are Anocracies More Prone to Civil War?" *Political Research Quarterly* 63, no. 4 (2010): 747–759.

[56] Havard Hegre, Tanja Ellingsen, Scott Gates, and Nils Petter Gleditsch, "Toward a Democratic Civil Peace? Democracy, Political Change, and Civil War, 1816–1992," *American Political Science Review* 95, no. 1 (2001): 33.

[57] Elsewhere, this is referred to as economic opportunity.

[58] James D. Fearon, "Governance and Civil War Onset," *World Development Report 2011 Background Paper*, 31 August 2011, 14–16, https://openknowledge.worldbank.org/bitstream/handle/10986/9123/WDR2011_0002.pdf.

[59] World Bank, *World Governance Indicators*, http://info.worldbank.org/governance/wgi/index.aspx#home; PRS, International Country Risk Guide, https://www.prsgroup.com/about-us/our-two-methodologies/icrg.

[60] Barbara F. Walter, "Why Bad Governance Leads to Repeat Civil War," *Journal of Conflict Resolution* 59, no. 7 (2015): 1424–1272.

[61] Paul Collier and Nicholas Sambanis, "Understanding Civil War: A New Agenda," *Journal of Conflict Resolution* 46, no. 1 (2002): 3–12.

[62] Kristian Skrede Gleditsch, "Transnational Dimensions of Civil War," *Journal of Peace Research* 44, no. 3 (2007): 293–309.

[63] Sarah Zukerman Daly, "Organizational Legacies of Violence: Conditions Favoring Insurgency Onset in Colombia, 1964–1984," *Journal of Peace Research* 49, no. 3 (2012): 473–491.

[64] J. Michael Quinn, T. David Mason, and Mehmet Gurses, "Sustaining the Peace: Determinants of Civil War Recurrence," *International Interactions* 33, no. 2 (2007): 167–193.

[65] Roy Licklider, "The Consequences of Negotiated Settlements in Civil Wars, 1945–1993," *American Political Science Review* 89, no. 3 (1995): 681–690;

[66] Matthew Hoddie and Caroline Hartzell, "Civil War Settlements and the Implementation of Military Power-Sharing Arrangements," *Journal of Peace Research* 40, no. 3 (2003): 303–320; Virginia Page Fortna, "Does Peacekeeping Keep Peace? International Intervention and the Duration of Peace After Civil War," *International Studies Quarterly* 48, no. 2 (2004): 269–292; Monica Duffy Toft, *Securing the Peace: The Durable Settlement of Civil Wars* (Princeton, NJ: Princeton University Press, 2009).

[67] Idean Salehyan and Kristian Skrede Gleditsch, "Refugees and the Spread of Civil War," *International Organization* 60, no. 2 (2006): 335–366.

[68] Seraina Rüegger, "Conflict Actors in Motion: Refugees, Rebels and Ethnic Groups" (PhD dissertation, ETH Zürich, 2013), https://icr.ethz.ch/publications/conflict-actors-in-motion/.

[69] Peter Waldmann, "Is there a Culture of Violence in Colombia?" *Terrorism and Political Violence* 19, no. 4 (2007): 593–609.

[70] However, as of 2010, more than half of the world's population lived in urban environments. That number is projected to steadily increase over the coming decades. As a result, future analysis of the effects of terrain on political violence and instability should also consider urban environments.

[71] Stathis N. Kalyvas and Laia Balcells, "International System and Technologies of Rebellion: How the End of the Cold War Shaped Internal Conflict," *American Political Science Review* 104, no. 3 (2010): 415–429.

[72] Most researchers use mountains (or slope elevation) and forests as a proxy for rough terrain. Little attention has been paid to other topographical features that impede government access or surveillance, such as swamps.

[73] See "Global Health Observatory: Urban Population Growth," World Health Organization, accessed February 3, 2014, http://www.who.int/gho/urban_health/situation_trends/urban_population_growth_text/en.

[74] Nathan Bos, "Underlying Causes of Violence," in Human Factors Considerations of Undergrounds in Insurgencies, ed. Nathan Bos, 2nd ed. (Ft. Bragg, NC: USASOC, 2012), 27.

[75] Håvard Hegre and Nicholas Sambanis, "Sensitivity Analysis of Empirical Results on Civil War Onset," *Journal of Conflict Resolution* 50, no. 4 (2006), 508–535; Paul Collier and Anke Hoeffler, "Greed and Grievance in Civil War," *Oxford Economic Papers* 56, no. 4 (June, 2004); James D. Fearon and David D. Laitin, "Ethnicity, Insurgency, and Civil War," *The American Political Science Review* 97, no. 1 (February 2003), 75–90.

[76] Jan Ketil Rød, Wenche Larsen, and Nels Petter Gleditsch, "Foliage and Fighting: Forest Resources and the Onset, Duration, and Location of Civil War," *Political Geography* 27, no. 7 (2008), 761–782.

[77] Raymond L. Bryant, "Shifting the Cultivator: The Politics of Teak Regeneration in Colonial Burma," *Modern Asian Studies* 28, no. 2 (1994), 225–250.

[78] Halvard Buhaug and Jan Ketil Rød, "Local Determinants of African Civil Wars, 1970–2001," *Political Geography* 25, no. 3 (2006), 315–335.

[79] Clionadh Raleigh, "Seeing the Forest for the Trees: Does Physical Geography Affect a State's Conflict Risk?" *International Interactions: Empirical and Theoretical Research in International Relations* 36, no. 4 (2010): 384–410.

[80] Halvard Buhaug and Scott Gates, "The Geography of Civil War," *Journal of Peace Research* 39, no.4 (2002): 417–433.

[81] Charles Butcher, "'Capital Punishment': Bargaining and the Geography of Civil War," *Journal of Peace Research* 52, no. 2 (2015): 171–186.

[82] Andreas Foro Tollefsen and Halvard Buhaug, "Insurgency and Inaccessibility," *International Studies Review* 17, no. 1 (2015): 6–25.

[83] Kenneth E. Boulding, *Conflict and Defense: A General Theory* (New York: Harper, 1962).

[84] Halvard Buhaug, "Dude, Where's My Conflict?: LSG, Relative Strength, and the Location of Civil War," *Conflict Management and Peace Science* 20, no. 10 (2009): 1–22.

[85] Sidney Tarrow, *Power in Movement: Social Movements and Contentious Politics* (Cambridge, UK: Cambridge University Press, 1998), 72.

[86] Ibid., 71.

[87] Ibid, 71–72; Doug McAdam, *Political Process and the Development of Black Insurgency, 1930–1970* (Chicago and London: University of Chicago Press, 1982), viii–ix.

[88] McAdam, *Political Process*, 6–11.

[89] Ibid., 40–42.

[90] Hanspeter Kriesi, "Political Context and Opportunity," in *The Blackwell Companion to Social Movements*, David A. Snow, Sarah A. Soule, and Hanspeter Kriesi, ed. (Malden, MA: Blackwell Publishing, 2007), 67-90.

[91] Charles Tilly and Sidney Tarrow, *Contentious Politics* (New York, NY: Oxford University Press, 2007), 55–57.

[92] Tilly and Tarrow note that the language describing political opportunities can be misleading. Sometimes, the shifts important for explaining outcomes are due to threats, not opportunities presented by expanded opportunities to participate in legitimate channel. In some cases, state repression can galvanize resistance, particularly violent resistance

[93] Ibid., 55–57.

[94] Katherine Raley Burnett, Christopher Cardona, Jesse Kirkpatrick, Sanaz Mirzaei, and Summer Newton, *Case Studies in Insurgency and Revolutionary Warfare: Colombia (1964–2009)* (Fort Bragg, NC: USASOC, 2012), 140.

[95] Kriesi, "Political Context and Opportunity," 67–90.

[96] Various internal security forces, however, opened fire on protesters and used other repressive measures to disperse crowds on Mubarak's order, but they were quickly overwhelmed by the sheer mass of protesters. Mubarak, like many other authoritarian leaders, used a number of so-called coup-proofing strategies to configure security institutions in such a way that they were in competition with one another for power and resources, rather than targeting the regime. Mubarak implemented two coup-proofing strategies: establishing parallel security institutions and distributing patronage and material resources. Mubarak's regime established several internal security forces outside the military, the Central Security Forces (CSF), the State Security Investigations (SSI), and the GIS. The CSF was tasked with quelling domestic opposition and checking the military, while the SSI was a secret police force also responsible for ensuring domestic security. Meanwhile, the GIS was charged with monitoring the Ministry of the Interior, which housed the CSF and the SSI, and monitoring the regular army. The overlapping mandates and jurisdictions encouraged intense competition for influence among the organizations, distracting them from targeting the regime itself. Moreover, Mubarak used significant patronage, particularly within the military, to cement loyalty. The military had the state's best salaries, housing, health care, and access to lucrative contracts in the private sector. Some estimate that the military's economic fortunes accounted for about 40 percent of Egypt's total economy. The coup-proofing strategies worked to drive a significant wedge between the military and the internal security forces, exacerbated when Mubarak's investment in the Ministry of the Interior eventually overtook the budget of the Ministry of the Defense.

[97] Michael Makara, "Coup-Proofing, Military Defection, and the Arab Spring," *Democracy and Security* 9, no. 4 (2013): 334–359.

[98] Holger Albrecht, "Does Coup-proofing Work? Political–military Relations in Authoritarian Regimes Amid the Arab Uprisings," *Mediterranean Politics* 20, no. 1 (2015): 36–54.

[99] Makara, "Coup-Proofing," 334–359.

[100] Ibid.

[101] Erica Chenoweth and Maria J. Stephan, *Why Civil Resistance Works: The Strategic Logic of Nonviolent Conflict* (New York: Columbia University Press, 2011), 46–50.

[102] George Lawson, "Revolution, Nonviolence, and the Arab Uprisings," *Mobilization: An International Quarterly* 20, no. 4 (2015): 453–470.

[103] Zeev Maoz and Belgin San-Akca, "Rivalry and State Support of Non-State Armed Groups (NAGs), 1946–2001." *International Studies Quarterly* 56, no. 4 (2012): 720–734. Idean Salehyan, *Rebels Without Borders: Transnational Insurgencies in World Politics* (Ithaca, NY. Cornell University Press, 2009).

[104] Idean Salehyan, "The Delegation of War to Rebel Organizations," *Journal of Conflict Resolution* 54, no. 3 (2010): 493–515

[105] Gerard Prunier, *Africa's World War: Congo, the Rwandan Genocide, and the Making of a Continental Catastrophe* (Oxford, UK: Oxford University Press, 2009).

[106] Idean Salehyan, *Rebels Without Borders: Transnational Insurgencies in World Politics* (Ithaca, NY. Cornell University Press, 2009).

[107] Kristian Skrede Gleditsch, Idean Salehyan and Kenneth Schultz, "Fighting at Home, Fighting Abroad: How Civil Wars Lead to International Disputes," *Journal of Conflict Resolution* 52, no. 4 (2007): 479–506.

[108] Halvard Buhaug and Kristian Skrede Gleditsch. "Contagion or Confusion? Why Conflicts Cluster in Space," *International Studies Quarterly* 52, no. 2 (2008): 215–233.

[109] Christopher Linebarger. "Civil War Diffusion and the Emergence of Militant Groups, 1960-2001," *International Interactions* 41 no. 3 (2015):583–600. Kurt Weyland, "The Diffusion of Revolution: 1848 in Europe and Latin America," *International Organization* 63, no. 3 (2009):391–423.

[110] Lars-Erik Cederman, Luc Girardin, and Kristian Skrede Gleditsch, "Ethnonationalist Triads: Assessing the Influence of Kin Groups on Civil Wars." *World Politics* 61 no. 3 (2009):403–437. Kristian Skrede Gleditsch, "Transnational Dimensions of Civil War," *Journal of Peace Research* 44, no. 3 (2007): 293–309.

[111] Idean Salehyan and Kristian Skrede Gleditsch, "Refugees and the Spread of Civil War," *International Organization* 60, no. 2 (2006): 335–366

[112] Sarah Lischer, *Dangerous Sanctuaries: Refugee Camps, Civil War, and the Dilemmas of Humanitarian Aid* (Ithaca, NY: Cornell University Press, 2006).

[113] David Cunningham, "Blocking Resolution: How External States can Prolong Civil Wars." *Journal of Peace Research* 47, no. 2 (2010): 115–127.

CHAPTER 3.
WHY DO SOME PEOPLE AND GROUPS MOBILIZE?

Resistance movements, whether violent or nonviolent, need people to participate.[1] The process through which individuals are brought into the fold goes by many names. Those analyzing resistance from the Special Forces's perspective focus on recruitment tactics and techniques. Most often, the Special Forces are concerned with recruitment in undergrounds and insurgent groups with the intent of distilling best practices. Academics in the social sciences, by contrast, are motivated by the search for an explanation of why certain people, at particular times, take part in collective action.

Collective action is a distinct form of social action because of who is involved and how individuals are involved. It describes action undertaken together by ordinary people or civilians acting separately from government in "confrontation with opponents, elites, or authorities."[2] However, collective action is regarded as the observable manifestation of a broader process, mobilization, which prepares people to take part in collective action.[3] Mobilization is "a dynamic, multistage process, not a singular event or discrete decision."[4] The explanation of mobilization is regarded as among the most important questions in the science of resistance because it requires sophisticated social solutions to occur. It requires bringing together people with different identities, interests, and locations to act in concert during sustained campaigns rife with uncertainty. This chapter discusses different theories regarding the barriers to mobilization and the mechanisms through which social solutions to those challenges occur.

The dominant theoretical framework for researching mobilization processes in the social sciences is Mancur Olson's free-rider problem. According to his theory, because resistance movements produce public goals when they are successful, individuals should logically choose nonparticipation because they can enjoy the benefits of those public goods without incurring the high costs of participation. This dilemma gives rise to the free-rider problem, whereby collective action requires a great deal of explanation.

In turn, the theory of selective incentive postulates that the leaders of resistance movements overcome the free-rider problem by offering additional incentives. The selective incentives, whether in terms of loot, natural resources, or attractive pay, are powerful financial motivation to convince otherwise reluctant recruits to join the group. The selective incentives theory is related to the greed versus grievance debate, whereby some argue that mobilization into resistance movements is based more on personal greed than any longstanding issues of grievance or injustice.

Since these theories emerged decades ago, others have posited explanations that fall outside these parameters. The theories are supplementary explanations to rationalist assumptions of the free-rider problem and selective incentives. Several relate to the role that social networks, affiliative factors, and emotions play in recruitment or mobilization. Other explanations for mobilization rely on the study of psychological risk factors or ideology.

This chapter also discusses micromobilization processes. As any practitioner of resistance knows, civil wars or insurgencies are not static. In fact, they are highly dynamic processes with frequent shifts throughout their life cycle. When the shifts occur, they subsequently alter the preferences, interests, and objectives of the groups and individuals involved directly or peripherally with the resistance. To capture these shifts and the impact on mobilization, researchers look at micro-level data that disaggregates political conflict into discrete events or individuals. One micromobilization model discussed here, the control-collaboration model, looks at how mobilization, or collaboration, is impacted by levels of territorial control exercised by the insurgents or an incumbent regime. When territory is contested, collaboration or recruitment is more likely because the raised costs of nonparticipation (or failing to choose sides) means near certain death. In these circumstances, violence and fear, not ideological commitments or other non-rationalist explanations, dominate the decision-making process.

Although the models presented in this chapter appear to be contradictory, they are in fact complementary. Numerous logic frameworks operate in a resistance cycle. One method of breaking down a resistance cycle is through conceptualizing different phases or stages. This chapter reviews how different logics of participation operate in the phases of resistance established by previous ARIS work, including the preliminary, incipient, crisis, and institutionalization phases. In the incipient and crisis phases, recruitment occurs in clandestine or underground organizations. Because of the high security risks and inability to offer selective incentives, in this phase, resistance movements rely on high levels of group identity, affiliative bonds, and identification between the individual and his or her recruiter. In addition, in these uncertain conditions, social networks are also a powerful mobilizing tool. These phases also tend to attract certain types of individuals called early risers or mobilizers that perceive less risk in resistance than others.

Although some of the findings are contested, recent research has demonstrated the effectiveness of nonviolent campaigns in achieving various political goals. In part, the success of the movement is theorized to hinge on the mobilization advantages of nonviolent campaigns. It could be that the campaigns are more likely to generate mass-based participation due to the lower physical, moral, informational, and commitment barriers to mobilization. Increased participation, in turn, may secure advantages for these campaigns. When participants are from numerous sectors of society, it is more difficult to isolate them as an aberration. However, it is not clear that nonviolent strategies always leads to mass mobilization. It may be that resistance leaders turn to violence to compensate for low participation. Finally, nonviolent campaigns might be more likely to obtain allies from abroad who offer crucial support in their struggle.

THE PUZZLE OF PARTICIPATION IN COLLECTIVE ACTION

Researchers are attracted to the question of mobilization because collective action requires leaders and visionaries of a resistance movement to overcome significant challenges. The challenges lie both in mobilizing a sympathetic population and in mounting an opposition against an incumbent regime that has ready access to the resources and advantages a resistance movement lacks. Indeed, Bowyer Bell accused resistance movements of being "criminally optimistic" about their chances of success.[5] As such, resistance movements use a variety of tactics and resources to overcome the barriers to participation.[6] The bulk of this chapter outlines those barriers and the myriad of explanations for how resistance leaders have successfully recruited others to their cause.

Before turning to the explanations, however, it is necessary to introduce the fundamental problem of collective action and what makes it so difficult, or not, to attract individuals to organizations pursuing group interests. This is an enduring puzzle that has fueled a great deal of social science research on resistance. Chapter 2 explored the structural explanations, such as political exclusion, poor socioeconomic development, and ethnic discrimination, which are often cited as motivations for participation. These explanations rely on large-scale structures, such as levels of state economic development or government policies of political exclusion, that motivate people to participate in resistance movements. As a result, the explanations do not include models of individual decision-making processes that prompt mobilization. The collective action framework presented later focuses nearly exclusively on this aspect of mobilization and is a powerful paradigm for explaining when and under what circumstances individuals are more likely to respond favorably to various recruitment strategies.

The collective action puzzle is most articulately expressed in the theory of economist Mancur Olson.[7, 8] In his seminal work, *The Logic of Collective Action*, first published in 1965, Olson challenged his fellow scholars studying the behavior of political groups. Most scholars, he argues, assume that individuals in groups, whether lobbying groups or labor unions, work seamlessly together in pursuit of the group's interests:

> In other words if the members of some group have a common interest or objective, and if they would all be better off if that objective were achieved, it has been thought to follow logically that the individuals in that group would, if they were rational and self-interested, act to achieve that objective.[9]

Assuming that individual behavior neatly dovetailed with group interests meant that there was not much about mobilization that needed to be explained. Using economic theories, Olson challenged the commonly held assumption; his logic exposes the many challenges surrounding mobilization processes. He argues that under most conditions, individual members of groups do *not* act in accordance with group interests, stressing that grievances, by themselves, are insufficient motivations to join resistance movements. Instead, individuals are expected to act within the group interest only if each is offered some sort of additional incentive outside the goods acquired through the achievement of group interests. In economic theory, which in turn heavily influenced how social scientists think about individual behavior, individuals are presumed to be rational actors in pursuit of their self-interest. This means that individual behavior and decision-making is based on calculations of utility or the calculations of costs and benefits. According to this logic, the actions with the greatest benefits and lowest cost are more preferable to those with higher costs and lower benefits. In economics, the self-interest of individuals or organizations is generally understood in terms of profit, while the term is more ambiguously applied in political science.

> *The problem of collective action asserts that it is not rational for individuals to act on behalf of a group's interest because it produces public, not private, goods.*
> *As a result, individuals have an incentive to let others accept the burdensome task of resistance, or free-ride, because even nonparticipants will enjoy any benefits the group produces.*

The critical point of failure in individual participation within political groups is the creation of public goods. Public goods are a class of goods that must be made available to everyone if they are made available at all. Services provided by the state are the best examples of public goods. The state provides national defense, police protection, and roads, among other goods, to its citizens. However, it is not possible, or at least practically feasible, to make roads or national defense available to just a few select persons.[10, 11] Instead, once roads or national defense are made available, the public as a whole has access to them. This means that even those who may not have paid or worked for the goods still can consume them. Public goods are also nonrival so that when one person enjoys national defense, it does not decrease the availability of national defense for other citizens.

The resulting dilemma is what Olson calls the free-rider problem, and he argues that it applies to political groups like resistance movements. Participating in a resistance movement, particularly one that is deemed illegal by the state or uses tactics deemed illegal by the state, such as violence, is a costly endeavor. Recruits oftentimes risk arrest, torture, or even death at the hands of state security forces or paramilitaries. Moreover, the benefits that resistance movements produce when they are successful are public goods. If a resistance movement persuades an oppressive government to enact policy changes advantageous to an ethnic group or government policy reform, for instance, all members of the affected group or society enjoy the benefits whether or not they personally participated in the movement. As a result, individuals have an incentive to let others take on the burdensome task of resistance, or free-ride.

After Olson first introduced the collective action problem, social scientists took pains to explain why some individuals and groups, but not others, seemingly defy the basic logic of the free-rider and take up arms or join mass nonviolent resistance movements. In the collective action paradigm, because the powerful logic renders mobilization such a challenge, its occurrence requires a solid theoretical and empirical explanation. As the following section discusses, some research emphasizes the additional incentives, called selective incentives, which make participation in resistance movements more attractive, while a smaller subsection of research supplements the logic of the rational individual with additional motivational considerations.

Selective Incentives in Collective Action

The collective action framework is the predominant framework researchers use to explain why some individuals or groups choose to participate in resistance movements.[12] The framework assumes that people are rational individuals who make decisions on whether or not to mobilize according to calculations of cost and benefit. If the cost of participation is too high and the benefits are too low or freely available, then the framework suggests that individuals will choose not to participate. However, with the levels of political resistance and violence across the world today, it is clear that many people find sufficient payoffs in participating in resistance movements.

Many researchers find the explanation to this puzzle in selective incentives or side payments that accrue to participants for their membership in a resistance organization. Samuel Popkin noted that a crucial mobilization strategy in Vietnam's rebellion against the French colonists relied on the provision of selective incentives. The colonial rulers paid peasants to participate in the defense of colonial rule, an action the peasants may not have taken without the selective incentives.[13] Similarly, in *The Rebel's Dilemma*, Mark Lichbach explains that offering selective incentives is one tactic rebel leaders use to attract followers who might not otherwise participate. Selective incentives are side payments that leaders provide to participants or that accrue to participants from belonging to the organization. Lichbach identifies a host of possible selective incentives, including land, money, loot, and positions of authority that can attract mobilization in a wide variety of resistance activities, from strikes to violent rebellion. With the addition of selective incentives, participants receive multiple payoffs because they will arguably still also enjoy the public goods produced by the organization. This means that each individual benefits from private goods as well as public goods.[14]

Interest in how selective incentives facilitate mobilization processes has motivated a debate between the relative role of greed or "grievance" in mobilization. Several researchers found that indicators outside of those typically associated with grievances are better predictors of the outbreak of civil war. They argued that rebel organizations were more likely to form when there were ample financial resources available for exploitation. In this case, financial resources were equated with the availability of natural resources. States that relied heavily on natural resources were more likely to experience civil war than even states with high levels of ethnic, political, or economic grievances.[15] However, the greed theory is misleading because it assumes the primary motivation of the participants is the accumulation of personal wealth. The emphasis on the negative value of greed overlooks other positive motivations, such as the need to support a family, that are also important to consider.[16]

Selective incentives are one explanation to the problem of collective action. Because individuals have incentives to free-ride, resistance leaders offer selective incentives or side payments to entice individuals to participate in collective action through the promise of personal reward.

It is difficult for an insurgency to sustain itself through legitimate businesses and voluntary contributions; therefore, many turn to more lucrative criminal enterprises. A highly valuable commodity that can be stolen and smuggled to finance a rebellion seems to make radical rebellion more likely. However, it may be important that these resources are located in rural areas, where insurgents can operate more freely and the need for safe long-distance transportation facilitates extortion. Natural resources prone to this sort of exploitation are hydrocarbons and gemstones, the so-called "blood diamonds."[17] The favorable recruitment environment also appears to contribute to conflict duration because civil wars in areas with high natural resource endowments are nearly twice as long.[18]

Social Networks and Affiliative Factors

Other explanations for collective action corrects the rationalist per-spective that assumes individuals are cost-benefit calculators. Instead, this class of explanations looks toward the individual need for affiliative bonds and the role of social networks in mobilization processes. Psychologist Abraham Maslow, who first developed the hierarchy of human needs, observed that humans require belonging and acceptance; they need to love and be loved by others.[19] When the needs remain unfulfilled, individuals may be more susceptible to join a resistance movement as participation fulfills those needs.[20] The needs are exacerbated by issues of social identity when individuals question their role in a new culture. They may seek out others with similar ethnic, social, and/or professional backgrounds.[21] In some cases, affiliative reasons for joining a resistance movement are more common than grievance-based or ideological motivations many attribute to mobilization processes. Individuals may be indoctrinated into a group's ideology only after being recruited into the resistance movement via their social networks.[22, 23]

Affiliative factors describe how the emotional needs for belonging and social inter-action can facilitate mobilization into resistance movements. When the needs remain unfulfilled, individuals may be more susceptible to joining a resistance movement after integrating into a social network for affiliative fulfillment that includes radicalized members.

Research on al-Qaeda illustrates the role social networks, as opposed to ideology, play in recruitment.[24] Biographies of four hundred al-Qaeda-affiliated radicals were compiled from trial transcripts, press accounts, academic publications, and corroborated Internet sources. Of that sample, 162 were from the Maghreb,[25] 132 were from core Arab countries, and 55 were from southeast Asia. Thirty-eight high-value individuals who were components of the al-Qaeda central staff are also further distinguished. The vast majority of these higher echelon members had secular, not religious, educational backgrounds. Egyptian Islamic militants who had been released from prison and who traveled to Afghanistan to fight against the Soviets composed most of the central staff. The central staff and Maghreb Arabs were upwardly mobile young men from cohesive middle-class families and possessed good technical skills.

Many of the research subjects spent time abroad. Their separation from traditional and cultural bonds prompted them to seek out social interaction with people of similar backgrounds. Importantly, many adopted al-Qaeda's ideological beliefs only after being integrated into social networks that included radicalized individuals. Al-Qaeda recruiters only accepted a small number of the interested individuals. Around 68 percent joined because of preexisting friendships with members, while another 20 percent joined because of familial ties with members. In 98 percent of the cases, social bonds preceded ideological commitment. There was no evidence of coercion or brainwashing; individuals simply acquired the beliefs of those around them after exposure. In each case, the individual joined via acquaintances, relatives, and imams, not through electronic or bureaucratic methods of recruitment.[51]

Social Networks and the 2011 Dar'a Uprising in Syria

In the aftermath of the Arab Spring uprisings in Egypt and Tunisia, protests against the authoritarian regime in Syria began to emerge. Syria was a particularly challenging location for formation of a resistance movement because the state's secret security forces, the dreaded Idarat al-Mukhabarat al-Amma, embedded extensive informant networks within the population. Many lived in fear of voicing any criticism of President Assad. Moreover, the father of sitting President Bashar alAssad, Hafez al-Assad, proved his willingness to brutally repress political opposition when he destroyed the city of Hama in 1982, the base of operations for a radical offshoot of the Egyptian Muslim Brotherhood.

Nevertheless, in 2011, the residents of Dar'a were the first to mobilize, acting as early risers, in the protests against the Assad regime that eventually escalated to full-scale civil war that killed hundreds of thousands and displaced millions more. What separated the backwoods Dar'a region from other cities and regions in Syria, enabling it to mobilize while others did not? Inhabitants of cities and regions throughout Syria held ample grievances against Assad and had witnessed the fall of the authoritarian regimes in Egypt and Tunisia.

Researcher Reinoud Leenders, who interviewed residents there in 2011, found that the answer likely lies in the region's particularly dense social networks.[26] The Dar'a region, paradoxically, was among the regime's staunchest supporters. Its support meant that Dar'awis held high-ranking positions in the Syrian government, military, and secret services, enjoying other benefits not widely available to others. The social networks in the region included clan affiliations, labor migration, cross-border traffic, and criminal groups. In addition to having a number of networks, Dar'awis were miscible, or highly interconnected with one another.

The uprising began when security forces repressed minor resistance activity, including graffiti writing on schools and police stations that mimicked slogans heard in Egypt and elsewhere. Tensions rose after the police arrested a number of schoolboys, aged ten to fifteen years, subjecting them to the same harsh treatment adult dissidents suffered. Local clan leaders and members of parliament pleaded for the release of the boys but were rebuffed. A series of antiregime protests and demands for the boys' release escalated into violence after police shot and killed tens of protesters. By April 2011, Dar'a was under military siege by regular army units, snipers, and tanks.

Dar'awis continued to mobilize in the streets despite the danger to themselves and their families. The social networks in Dar'a facilitated mobilization in a number of ways. First, they reportedly offered residents a sense of solidarity and shared risk. The social networks also helped in less obvious ways. Residents reported feeling pressured by their peers to mobilize. Also, as one resident noted, when your clan leader tells you to do something, you do it. Clan pressure to join the resistance was also notable among former influential regime loyalists, some of whom felt compelled to criticize the Assad regime for the violence or resign their positions in protest. As the protests continued, additional motivations to join emerged. The violence took the lives of an estimated 632 Dar'awis in 2011, prompting others to join after experiencing deep personal loss at the hands of security forces and the Assad regime.

Finally, illicit social networks provided skills and resources necessary for mobilization, even under the harsh conditions of repression. Labor migrants smuggled in satellite phones. Taxi and truck drives smuggled out video footage for YouTube and smuggled in needed medicines, food, and weapons. The criminal networks provided training in weapons, intelligence gathering, running safe houses, and establishing temporary hospitals to treat the wounded.

Dar'a's role as an early riser, facilitated by its dense web of social networks, was a powerful demonstration to others in Syria that the regime was not as all-powerful as others had assumed. The actions of the Dar'awis helped embolden others to take similar actions, even mimicking the neighborhood defense committees first established there. The regime's brutal repression of Dar'a helped to escalate the conflict from a nonviolent uprising to an armed insurrection.

Reinoud Leenders, "Collective Action and Mobilization in Dar'a: An Anatomy of the Onset of Syria's Popular Uprising," *Mobilization: An International Journal* 17, no. 4(2012): 419-4–34.

EMOTIONS AND IN-PROCESS INCENTIVES IN MOBILIZATION

The logic of the collective action framework stipulates that individual decision-making is focused on calculations of benefit based on future outcomes. However, other researchers find that individuals also join resistance movements because of in-process benefits, or those benefits that a person gains while participating. Most of the arguments for the attractiveness of in-process benefits highlight the emotional benefits gained from participation in a movement. Elisabeth Wood, using extensive interviews with participants and nonparticipants, attributes participation in the leftist insurgency in El Salvador (1979–1992) to emotions. Many individuals who eventually joined the armed insurgency had first started their activist careers in nonviolent resistance calling for economic and political reform. As government repression of these efforts increased, some individuals chose to take up arms to reassert "their dignity in the face of condescension, repression, and indifference."[27] The participants reported feeling a sense of pride and pleasure in standing up for their interests, a characteristic Wood calls "pleasure in agency."[28] In this regard, the process is valued as much as the outcome of resistance.

In-process benefits are the emotional benefits a person experiences while participating in a resistance movement that can serve as a motivation for joining and staying in a resistance movement.

Emotions, despite being associated with irrationality, are potent strategic tools leveraged by resistance leaders and counterinsurgent actors. Terrorist attacks, for instance, are intended to elicit emotional responses from targeted audiences to provoke certain retaliatory behaviors. Emotions are also effective at altering individual perceptions of risk. Emotional responses resulting from belonging in intimate social networks, holding mass meetings, activating collective identities, or shaming are an important part of decreasing an individual's sense of risk. When fear of repression is lessened, people are more likely to participate in collective action. Fear abatement through emotional responses, then, is particularly important in resistance movements that face a high risk of repression.

One important emotion for mobilization is moral indignation, often activated in injustice frames. Injustice frames are interpretations proffered by movement leaders that highlight how adversaries are actively bringing about suffering or harm to affected groups. When successful, the frames result in "hot cognition," or a recognition of inequitability melded with strong emotional content that "puts fire in the belly and iron in the soul."[29] Indignation is an emotion frequently associated with an individual's motivation to seek out protest groups.[30] Indignation results when moral expectations or principles are shocked in some fashion. There is a long list of possible expectations or principles that, when breached, are liable to produce feelings of indignation, including professional ethics, religious beliefs, community allegiances, feelings of physical security, or political ideologies.[31] Indignation may be an important component of recruitment when preexisting social networks for recruitment are absent, connecting movement leaders with strangers.[32] A strong sense of moral indignation can be a potent motivator for mobilization into collective action.

Injustice frames are interpretations proferred by movement leaders that highlight how adversaries are actively bringing about suffering or harm to affected groups. When successful, injustice frames help to ignite emotional responses, including "hot cognition" that facilitates participation or support with the group.

In the civil rights movements in the United States and East Germany, where the risk of repression were quite high, the movement leaders used a variety of techniques to alleviate crippling fear of economic reprisal, physical harassment, bodily injury, arrest, and even death at the hands of security forces. Alleviating the fear was crucial because it might have otherwise derailed the movements' efforts. A variety of encouragement mechanisms succeeded in both cases. One encouragement mechanism was the intimate social networks that formed among participants. The deep interpersonal relationships fostered a strong sense of solidarity among participants, raised the costs of dropping out, and shielded participants from outside pressure hostile to the movement's goals.[33] Security forces, meanwhile, also adjusted their behaviors to invoke greater fear responses among government opponents.[34] Conflicts are replete with this back-and-forth emotional and psychological warfare.

Humiliation

Humiliation and the consequent internal pressure for revenge is another emotional state conjectured to predispose individuals to join a resistance movement.[35] Experiences of grief are often accompanied by strong feelings of humiliation, which present a risk factor for the transition to violent tactics. The greater the degree to which a group is subjected to physical repression or torture or perceives itself to be humiliated by its enemies, the greater the risk that the group will take revenge through political violence.[36]

Comprising individuals' sense of their own dignity also provides a motivation to defend everyone within the in-group. Revenge is an emotion that is likely rooted in the instinct to punish transgressors who violate the rules of early groups necessary for survival. It is a motivator that often serves not only the goals of a vengeful individual but also the goals of the group.[37] The humiliation of political opponents can create an environment that leads to more violent behavior, aggravating a conflict or escalating tactics to include violence. Political, ideological, and religious narratives may mediate between the collective identity and personal misery from humiliation, but they may also reinforce a victimization identity that contributes to increased potential for violent behavior.[38]

Cultures in which humiliation is a frequent motivator for action are called cultures of honor.[39] In such cultures, there is an imperative among members to preserve honor by avenging slights, sometimes through the use of violence. Particularly for men, failing to retaliate for an insult, attack, or property encroachment can be seen as a serious threat to the individual's honor and reputation. Examples of these cultures exist all across the world, including rural areas in the American South, among American urban gangs, and some populations in the Middle East.[40] They are characterized by high homicide rates, cycles of retaliation, and long-running feuds.[41] Research indicates that cultures of honor are particularly associated with group low-status driven by lower socioeconomic development, poorer education, and disparaging stereotypes of the group in mainstream culture.[42]

Humiliation, particularly linked to notions of masculinity, played a significant role in Islamist political violence. Roxanne Euben observes that masculine humiliation vis-à-vis the West is frequently referenced in Islamist discourse. Humiliation is depicted as "an imposition of impotence on Islam/Muslims by those with greater and undeserved power."[43] The poor socioeconomic development in most Muslim-majority countries, legacies of colonialism, the Israeli occupation, and cultural hybridization are all conditions that can make it challenging for Muslim men in these regions to feel like they are performing culturally prescribed masculine roles, whether entering into marriage, protecting women and children, or defending the umma.[44] The associated impotence is interpreted as a violation of natural gender and sexual norms necessary for the appropriate religious-based social and moral hierarchy.

Retaliatory action prompted by masculine humiliation is a reassertion of morally appropriate gender roles. The humiliation emphasized in Islamist discourse, exemplified especially in the work of Abdullah Azzam, the architect of the Afghan jihad and a former mentor to Osama bin Laden, brings attention to the assault on the Muslim male's capacity to defend the umma and his family. Azzam's exhortations to the would-be Afghan mujahideen highlight the plight of young Muslim men in Afghanistan unable to protect Afghan women, or their children, from being raped and killed. In Azzam's work, and in the propaganda of al-Shabab and ISIS, women often feature as the symbol of humiliation. The violation of Muslims is also a violation of the Muslim masculinity where "[women's] bodies are the primary battleground for the humiliation of Islam, [women's] voices are a chorus of praise for such acts, and [women's] virginity is the eschatological reward for those who die humiliating the enemies of Islam."[45]

Similarly, a clan leader and local member of parliament in Dar'a, the Syrian region attributed with jump-starting the Arab Spring in Syria, reported experiencing masculine humiliation in his interview with local police. His story was told and retold throughout Dar'a, igniting a fuming response from local Dar'awis that led to a military siege in the city. The clan leader reportedly met with a local intelligence chief to plead for the release of several young schoolchildren imprisoned for antiregime graffiti. The intelligence chief reportedly told the clan leader, "Your people either accept things as they are, or you bring their women to me and I make them conceive some new kids." The inflammatory interview also took place after officials detained two local women, beating them and shaving their heads. Residents reported rage as the actions "breached the honor of their women." Both events, and the rhetoric of humiliation that spread through the town, became a central feature of a nonviolent protest "against indignity, rather than submission."[46]

The participation of women, by contrast, in violent groups is less studied. More often, research on women in conflict emphasizes their roles as victims of sexual violence or their vulnerability to indiscriminate violence against civilians. Research focusing on women as agents, not just victims, in conflict is more rare. However, it is evident that women take up arms less often than their male counterparts. In a study of seventy insurgent groups, women participated in either support or combatant roles in 60 percent of the groups. While some of the women were coerced into participation, more joined these groups voluntarily.[47] Some of the restrictions on women's participation in insurgent groups is likely due to ideology. Groups that rely on leftist Marxist ideology have higher rates of female participation than groups with Islamist discourse. One key difference between the ideologies relates to their position on gender norms and hierarchy. Marxist ideologies focus on dismantling existing gender norms and social hierarchies, while Islamists seek to protect them against encroachment.[48]

Women join insurgencies for reasons similar to men. They join to support a political or social cause and seek out combat roles because of the prestige associated with such roles.[49] However, women combatants of one insurgent group, the Farabundo Marti National Liberation Front (FMLN), also used gendered narratives to describe their motivations, including the indiscriminate, brutal attacks of El Salvadoran security forces against women, children, and the elderly. By contrast, the women portrayed the FMLN as the protector of the vulnerable.[50] A broader study of numerous insurgent groups confirmed that women participate in combat roles when there are threats to human security in their society. These women were less likely to be attracted to selective incentives or driven by socioeconomic grievances.[51]

PSYCHOLOGICAL RISK FACTORS

Some research on mobilization factors look at whether individual psychological conditions predispose some people to participation in resistance movements, particularly ones that use violence.[52] When social scientists first began to systematically study resistance after the end of the Second World War, researchers evidenced significant biases against the subjects of their research. Many, having directly or indirectly witnessed participation in communist and Nazi political movements and parties, equated participation in even relatively benign resistance movements with abnormality or mental disorder. In research related to social movements and political violence, the participants were assumed to partake in large-scale irrationality as the mood of gathered crowds degenerated into mob mentality. Participants were also believed to suffer from mental health disorders, feeling a deep sense of isolation or alienation mitigated only by their participation in forms of collective action that brought a sense of belonging back to the individual. The research focus neglected the political, social, or economic circumstances that fostered the resistance in favor of introducing resistance as an individual psychological phenomenon. It was believed that participation in resistance movements relieved individual mental wounds rather than widespread societal grievances. Resistance itself was regarded as irrational or a form of abnormal politics that had little hope of succeeding against the significant powers of the state. The research on resistance gained more subtlety and theoretical sophistication as more researchers sympathized with the social movements emerging in the era or participated themselves.

Some similar biases are still evident in laymen explanations of resistance, especially resistance associated with indiscriminate violence or terrorist tactics. Media commentators and even some scholars describe the perpetrators' acts and mental health to suggest that they are likely to have mental disorders.[53] However, most terrorists or insurgents are psychologically normal. Additionally, there is no psychological or demographic profile that indicates a predisposition toward joining violent organizations.[54] No all-purpose terrorist profile has been discovered, nor does scientific evidence of a genetic role in the adoption of terrorist behaviors exist.[55] In general, radical organizations likely have a sufficient range of personality and cognitive profiles within their ranks that their members are indistinguishable from the surrounding population.[56] Furthermore, severely mentally ill people usually have difficulty fitting in within teams and larger organizations. This holds true whether the organization in question is a corporation or a resistance movement. It is more likely that recruiters in resistance movements weed out mentally unfit individuals to retain the effectiveness of the group and operational security.

IDEOLOGY

Georges Sorel, an astute political theorist who focused on the human experience of political violence, observed that "men who are participating in a great social movement always picture their coming action as a battle in which their cause is certain to triumph."[57, 58] Outside of selective incentives or other rational calculations, successful resistance movements rely on ideologies to help overcome challenges to mobilization. Ideologies are expressed in narratives that aid in recruitment, legitimacy, and support that resonates with a deep cultural, ethnic, or historical memory within a sympathetic population.

An ideology is a comprehensive set of interrelated beliefs, values, and norms. Every society shares commonly held cultural beliefs, including ideas, knowledge, lore, superstitions, myths, and legends. The beliefs in turn are associated with values or judgments of right or wrong that guide individual action. This code is reinforced through a system of rewards and punishments so that approved patterns of behavior, or norms, can discipline the behavior of the group.

An ideology is a comprehensive set of interrelated beliefs, values, and norms.

Because beliefs and values are only distantly related to concrete action in daily life, an interpretive process is essential to derive specific rules of behavior. Significant events that occurred in distant times are given symbolic meanings, and the group actively reinterprets these events to support its purpose, goals, and tactics. In doing so, the group may select certain concepts and adapt or distort them to justify specific forms of behavior. Where existing concepts conflict with current activities, the group may deny that a particular concept is relevant in a particular case.

Within organizations, certain rules specify desirable behavior and the consequences of not conforming. The rules are enforced by organizational rewards and punishments relevant to the objectives of the group. Normative standards are also enforced by surveillance of members. In established groups, many beliefs are based on authority; because the leaders of the group voice the beliefs, they are accepted as true. The extent of authority, however, may vary according to whether a group is hierarchically or non-hierarchically organized. When leaders control the dissemination of information to the members of an organization, they censor some information and approve other types of information. As a result, the group receives a restricted range of information, and group members tend to develop a set of common beliefs. Thus, in some cases, members need not be persuaded by argument, induced by reward, compelled by pressure, guided by past beliefs, or influenced by the opinions of other people; the restricted range of information to which they have access is sufficient to determine their beliefs.

Ideology is a powerful tool because it helps individuals to reduce uncertainty about how the world around them functions. Human beings dislike ambiguity and uncertainty in their social and physical environments. Uncertainty reduction theory holds that most people do not congregate into groups unless there is a motive to alleviate uncertainty.

This uncertainty arises from the inability to obtain confirmation of one's beliefs and attitudes from objective criteria or measurements. It is a source of stress that has cognitive, affective, and behavioral consequences that can inhibit decision-making and action.[59] While intolerance of uncertainty presents a cognitive risk factor for anxiety, there are no empirical data to support it as a risk factor for mobilization.[60] Nevertheless, ideologies provide individuals with a comprehensive conceptual framework through which to interpret the existing world and process novel stimuli, a potent insulation against uncertainty.

Individuals seek to give meaning and organization to unexplained events through generalized beliefs such as ideology. Common agreement on certain beliefs is also a social solution that enables individuals to operate collectively toward a desired goal. Movement leaders can interpret situations in terms of the group's beliefs or ideology, translating abstract, ideological beliefs into specific, concrete collective actions.

Qutbism: Ideology of the Modern Global Salafist Jihad

The modern global Salafist developed over a series of phases. The first phase was the largely clandestine Islamist movement in Egypt, with the Egyptian Islamic Jihad (EIJ) and the Egyptian Islamic Group (EIG) being the more prominent of the groups that emerged from the Muslim Brotherhood. The second phase was the mass mobilization of Muslim youth to participate in the defensive jihad in support of the Afghan Mujahidin's resistance to Soviet occupation. The ongoing phase resides in the al-Qaeda resistance movement, a loosely affiliated group of networks linked by a common ideology.[61, 62]

Islamic theorist Sayyid Qutb provides the ideological foundation of the late twentieth-century global Salafist movement. The concept of Salafism has been employed by Sunni theologians since at least the fifth Muslim generation to differentiate the creed of the first three generations following Muhammad from subsequent variations in the Muslim belief system. Salafists view the first three generations as an eternal model for all succeeding Muslim generations, especially in their beliefs and methodology of understanding the texts and also in their method of worship, mannerisms, morality, piety, and conduct. Salafists place a particular emphasis on monotheism (*tawhid*). They also reject Islamic speculative theology (*kalam*), which involves the use of discourse and debate in the development of the Islamic creed. They believe that Islam's decline after the early generations results from religious innovations (*bid'ah*) and from an abandoning of pure Islamic teachings. An Islamic revival will only result through the emulation of the three early generations of Muslims and the purging of foreign influences from the religion.[63]

Qutb, an Egyptian, published his most famous work, *Milestones or Signposts along the Road*, in 1964. The book was banned shortly after its publication. His writings motivated disillusioned young Muslims who sought a more active role in returning Egypt to the center of the Islamic world. Throughout the 1950s and 1960s, political Islam served as an intellectual and ideological counterweight to the uniquely Egyptian blend of Arab nationalism and socialism espoused by the government of President Gamal Abdel Nasser.

In *Milestones*, Qutb presented his comprehensive view of the cosmos and the role of the Muslim community in it. He believed much of the world to exist in a state of ignorance (*jahiliyyah*) and that submitting to Islam was a political, social, psychological, and spiritual liberation. The way to bring about this freedom was for a revolutionary vanguard, or jihad, to combat ignorance through preaching and through abolishing the organizations and authorities of all un-Islamic systems. This movement would spread across the Islamic homeland and ultimately throughout the world, with all attaining freedom through submission to Islam.[36] Both EIJ and EIG emerged from this Salafist community.

Mohammed Abd al-Salam Farraj, the founder of EIJ, wrote *The Neglected Duty or the Absent Obligation*, which has been retrospectively classified as the manifesto and operational manual of the EIJ. In this text, Farraj began by stating, "Jihad for God's cause . . . has been neglected by the Ulema of this age."[64] He expanded the interpretation of jihad to include the violent struggle that is a duty (*fard al-ayn*) incumbent on all Muslims as it was the only way to reinstate a truly Islamic society. He incorporated Qutb's *jahiliyyah* but extended the charge to modern apostate Islamic rulers. Farraj advocated establishing an Islamic vanguard, or an elite cadre of pious Muslims. The vanguard was to serve as a model for elites in other Muslim nations to emulate. He made the initial classification of the "near enemy" (the impious Egyptian government) and the "far enemy" (Israel), subordinating all Islamic goals to the fight against local apostates.[65] The EIJ also granted itself the political and religious authority to declare all those who did not meet its requirements for piety essentially non-Muslim.[66] Other EIJ leaders, Sayyed Imam alSharif (Dr. Fadl) and Ayman Al-Zawahiri, also emphasized the importance of the Qutbist ideology. While in Afghanistan and Pakistan, Zawahiri developed a close working relationship with Osama bin Laden.[67]

The Iraqi annexation of Kuwait in 1990 and the subsequent deployment of US and coalition forces to the region resulted in an update to the Salafist insurgent ideology. Bin Laden saw the stationing of US troops in Saudi Arabia, home to Islam's holiest sites, as an unforgiveable offense. The Americans' increasing role in the region erased the residual goodwill it had earned by abstaining from colonization and supporting the Afghan resistance. In 1996, al-Qaeda announced its intention to expel foreign troops and interests from Muslim territory. Bin Laden issued a fatwa entitled "Declaration of War against Americans Occupying the Land of the Two Holy Places," a public declaration of war against the US and its allies.

He began to refocus the organization's resources toward large-scale psychological operations. The fatwa represented an overall shift in focus from the near enemy, Muslim apostate governments, to the far enemy, the United States.[68]

In June 2001, although difficult to distinguish for years, al-Qaeda and EIJ merged, forming Qaeda al-Jihad.[69] Zawahiri was presumed to be the deputy to bin Laden and the leader of the EIJ. While the charismatic leadership of bin Laden was by now well known, al-Qaeda's strength and appeal did not lay solely in its sophisticated theological discourse; it was also apparent in its ability to comprehend, co-opt, and exploit modern grievances. This narrative combination resonated with extremists and moderates alike, regardless of whether they approved of the means by which al-Qaeda sought to accomplish its goals. Al-Qaeda's leadership was not composed of highly trained religious scholars, and its religious rhetoric was far from complex or nuanced, making it broadly accessible. The specific messages within the larger narrative rarely focused on citing authoritative texts but rather relied on the application of general religious or ethical principles to modern political and social problems.[70]

MICROMOBILIZATION PROCESSES

One of the shortcomings in research in the social sciences on mobilizations processes, including the previous discussions on the collective action framework and psychological approaches, is that resistance, or civil war, is treated as a static phenomenon. In fact, resistance movements, and the insurgencies or civil wars that sometimes result from resistance, are highly dynamic processes that shift over the history of the conflict. As the conflict progresses, actors' preferences and objectives are prone to significant shifts according to changes internal and external to the resistance movement.[71] Stathis Kalyvas and Matthew Kocher explain that civil wars are processes that generate incentives and constraint, not a situation in which "payoffs to participation derives almost exclusively from expectations about outcomes."[72] This means that the mobilization processes that occur in one time or place during a conflict may not accurately reflect the processes at a different time or place.

Efforts are under way to better understand how the dynamics of conflict can impact mobilization at a more discrete level. Some researchers began these efforts by disaggregating conflict into smaller, constituent parts and collecting data, called micro-level data, accordingly. The micro-level data contrasts with the data found in most of the large-n studies on the macrostructural conditions of civil war, discussed in chapter 2; the data in these large-n studies represent an entire country, rather than the geographical areas where the conflict is actually taking place. The large-n data are also given a value for the conditions over a year's time, rather than for discrete episodes or incidents when conflict events occur.

More recently, scholars attempted to collect more targeted data on conflicts. This includes subnational data, or data that captures conditions in specific geographic locations where violence is actually taking place.[73] Other micro-level data break down conflict into specific incidents or individuals. This type of research is especially important for explaining mobilization at the individual level or why some individuals are successfully recruited while others are not. The targeted data, although difficult to collect, are better suited to producing evidence that is actionable or holds policy implications by focusing on how local dynamics impact mobilization.[74] However, the study of micromobilization, as it is called, is a relatively new field, and researchers have not yet developed a systematic set of concepts, measures, and research questions that enable a significant accumulation of knowledge on the subject.[75]

Micromobilization processes are discrete components of larger mobilization processes that take place over the course of conflict. Micromobilization uses micro-level data, whether at the level of the individual, geographic region, or phase in a conflict to better explain how mobilization occurs.

The research points to the difficulty in developing a master motivation theory that accounts for motivations for joining resistance movements in all times and places. Instead, research identifies how the motivations for mobilization might change according to shifting dynamics within the conflict.[76] This means that the rival explanations presented in this chapter, such as those emphasizing selective incentives, social and affiliative factors, and emotions, are likely not rival models. Rather, each explanation is likely to be predominant as the dynamics of the conflict shift over time. Additional work in the ARIS research program identifies different phases of resistance movements. Throughout its life cycle, a resistance movement might shift from one phase to the next, not necessarily in a linear fashion if the movement experiences setbacks. The motivation for joining at each phase, and thus the most effective recruitment tactics or strategies to deploy, require refinement accordingly. The emphasis on resistance or conflict as a process of complex interactions between involved actors over time, and what that means for the science of resistance, is a promising area of research for future ARIS studies.

In the micro-level or micromobilization research, there are several commonly studied variables or mechanisms that drive the dynamic processes. The first, the control-collaboration model, focuses on the interactions between civilians and armed actors. It is primarily used to explain variations in violence against civilians during an insurgency, a research effort of great concern to policy makers. However, the control-collaboration model, like the collective action framework on which it is based, is a powerful tool for thinking about under which conditions, and why, individuals are likely to join a violent organization. The next involves looking at mobilization, and thus recruitment, in different phases of an insurgency. The emphasis is on how mobilization during clandestine or underground phases differs with mass mobilization in later phases.

Control-Collaboration Model

The control-collaboration model incorporates the interactions between civilians and armed actors to enable a better understanding of mobilization processes. The model applies the same basic logic of the collective action framework but focuses on how the dynamics of violence, irregular warfare, and territorial control impact the mobilization preferences of individuals in affected communities. As a result, it is a nuanced, sophisticated model for thinking about how the dynamics of violence are liable to condition individual decision-making on whether to mobilize into an armed insurgency. It has powerful ramifications for thinking not only about mobilization but also about influential counterinsurgency theories that emphasize population attitudinal changes, "winning hearts and minds" under similar conditions of violence.

The control-collaboration model incorporates the interactions between civilians and armed actors to enable a better understanding of mobilization processes. The model applies the same basic logic of the collective action framework but focuses on how the dynamics of violence, irregular warfare, and territorial control impact the mobilization preferences of individuals in affected communities.

The control-collaboration model emphasizes the high cost of mobilization in violent conflicts. By contrast, the collective action framework incorporates individual expectations of public-goods payoff to explain decisions regarding mobilization. The emphasis in the control-collaboration model is the high cost of resistance because it theorizes that decision-making under conditions of violence is distinct as armed insurgencies are characterized by patterns of violence that target non-participants. In the collective action framework, the free-rider problem that introduces a barrier to mobilization is captured by considering both the high costs and the public-goods benefits of mobilization. Moreover, the benefits gained from collective actions are public goods, which means that they are nonexcludable and nonrival. The goods are nonexcludable because once the benefits, such as reformed policies for a politically excluded group, are available, everyone in the affected group can take advantage of the reforms regardless of whether or not they participated. In addition, the goods are also nondivisible because one person's enjoyment of them does not mean that anyone else has less of them to enjoy. However, most insurgencies are noxiously durable, and few succeed in achieving their stated goals, making it more difficult to recruit individuals based on the hope of future goods, particularly public ones.

The control-collaboration model, which is suited specifically to explain mobilization in violent conflicts, emphasizes the high cost of mobilization over this issue of public goods. In nonviolent collective action mobilization, recruitment might only require asking an acquaintance to sign a petition, a mostly costless contribution. On the other hand, recruitment for an armed insurgency requires persuading an individual to risk arrest, imprisonment, or death, a much more difficult task—high costs indeed.[77]

The patterns of violence, particularly the targeting of civilians, turn the logic of the collective action problem on its head. In this case, free-riding, or opting to not participate in collective action, is no longer free because nonparticipants are exposed to great physical insecurity even if they do not pick up a gun; that is, nonparticipation, just like mobilization, has a high cost because "war is very dangerous for non-rebels as well."[78] Sometimes, as will be shown later, the costs of nonparticipation are actually higher than those of participation. This means that under the conditions described in the model, the barriers to mobilize are not nearly as high as predicted by the collective action framework. Kalyvas and Kocher observe:

> If the collective action paradigm has been so dominant, it is because scholars have tended to overestimate the risks to rebel fighters or to underestimate the risks paid by non-participants—a result of limited attention to the dynamics of violence and of the tendency to impute preferences rather than investigate them empirically.[79]

The costly problem of nonparticipation posed by the peculiar patterns of violence in irregular warfare is best understood by contrasting it with (mostly) nonviolent resistance and conventional warfare. In conventional warfare, uniformed state armies are pitted against one another in sustained campaigns. Most of the violence associated with this type of warfare is concentrated in specific geographical areas, the so-called front lines. When civilians are exposed to this sort of violence, because the geographical area is a somewhat known factor, civilians can readily flee the area.[80] In this case, soldiers run the highest risk because they are directly engaged in warfare and, unlike civilians, have little to no option to flee the violence. Moreover, in contrast to conditions of irregular warfare, a soldier's enemies are generally readily identifiable. Similarly, in nonviolent resistance events, such as mass protests or demonstrations, people have the option of retreating to safety by not participating in protests or riots that are likely to be met with government repression. Staying home significantly decreases the risk of being targeted by state violence.

The irregular warfare common to civil wars presents a much greater risk to civilian populations because of the identification problem.[81] The identification problem is a thorny issue experienced by state actors or military personnel deployed in counterinsurgency campaigns such as those in Iraq and Afghanistan. In short, in conditions of irregular warfare, insurgents are dispersed among the civilian population. It is exceedingly difficult for their opponents to identify which individuals are the insurgents and which are the innocent civilians. The identification problem makes countering insurgencies in civil war a battle often decided by intelligence operations.

When state forces have sufficient intelligence, they can rely on precision targeting of known armed insurgents or discriminate violence.

However, when armed actors lack adequate intelligence, they often rely on indiscriminate violence that targets whole categories of individuals. From the state's perspective, it has incentive to target civilians. It might indiscriminately target whole ethnic groups, particular genders and age groups (usually men of fighting age), or certain geographic areas, such as villages believed to house insurgents.[82] Violence is discriminate when it relies on individual culpability, and it is indiscriminate when it widens the parameters of culpability to mere association or collective guilt. In this manner, when states wield indiscriminate violence, the violence increases the risk for any members of the targeted category, regardless of their personal behavior (i.e., whether they participate or not).[83]

When indiscriminate repressive violence is the primary tool used by state forces combating an insurgency, civilians have ample motivation to join the armed insurgency, and these motivations do not always coincide with ideology, greed, or grievance. Instead, fear for one's physical safety is of paramount importance in the mobilization process. Joining an insurgency offers some protection against government or occupational forces. As indiscriminate counterinsurgent measures increase, the cost of *not* joining the insurgency begins to mount; in other words, nonparticipation is costly.[84]

In a study of demobilized FMLN fighters in El Salvador, an anecdote in which a guerrilla fighter describes her reasons for joining powerfully illustrates the logic of the control-collaboration model. When asked by her interviewer why she joined the FMLN after her village was indiscriminately attacked by the Salvadoran Armed Forces, she replied:

> Because we couldn't just ask to live. If it wasn't the guerrillas, it was the Armed Forces. Because you see, here, if I stay, the Armed Forces kill me. If I go where the Armed Forces are in control, the guerrillas will kill me. That's why I went. What's more, the Armed Forces had killed nearly all of our family, so I certainly couldn't follow them.[85]

Viterna refers to this type of motivation for mobilization the "reluctant guerrilla."[86]

In this regard, the control-collaboration model exposes the fallacy that all civilian support for an armed insurgency is indicative of individual sympathy for the organization. Instead, the absence of any viable alternatives produces "collaboration, irrespective of the level of personal satisfaction or lack thereof."[87] Observers to these dynamic mobilization processes, including the state forces, wrongly interpret the collaboration as a reflection of the legitimacy of the enemy. Once large-scale violence is under way, the patterns of violence characterizing civil war prompt affected civilians to conceal their preferences, whether those preferences are in favor of the insurgents, the state forces, or mere survival.

Under these conditions, an individual's observable behavior, such as collaboration, is a poor indicator indeed of his or her actual preferences. Violence, particularly indiscriminate violence against the civilian population, has enormous coercive power over those experiencing it, making it the single most important factor in recruitment or collaboration.[88] As David Stoll insightfully observes, "just because an insurgency grows rapidly does not mean that it represents popular aspirations and has broad popular support."[89] The levels of coercion make it difficult to engage in counterinsurgency campaigns that emphasize winning the hearts and minds of the civilian population as a key strategy in defeating armed insurgents. In a crucial sense, the hearts and minds of civilians are immaterial because civilians' actions will continue to be motivated by fear until the pattern of violence abate.

In the control-collaboration model, indiscriminate violence is more likely to occur according to varying levels of actors' control of territory in the disputed state.[90] Kalyvas describes five levels of territorial control on a continuum, with complete control by either the insurgents or the state forces on either end. In the middle of the continuum, neither the state nor the insurgents have full control of the territory in question. The remaining two levels describe either predominant territorial control by the state or insurgent forces. According to the model, the more territorial control an armed actor enjoys, the less likely the group is to resort to indiscriminate violence. As a result, indiscriminate violence is most likely in the middle of the continuum, where neither actor enjoys even a modicum of control but is heavily disputing control of the territory with its opponent. Selective violence is most likely to occur when an armed actor has predominant control of a territory, enabling it to gain access to information and intelligence that allow for more selective violence against its opponent.[91]

Resistance Phases and Mobilization

The preceding chapter discussed numerous theories regarding how mobilization processes unfold. These theories can help students of resistance movements understand why some individuals or groups become actively involved in resistance activities or some do not. Some of the theories, such as the rationalist collective action framework and the nonrationalist explanation provided in the role of social networks and affiliative bonds in mobilization, appear to be contradictory. However, as Macartan Humphreys and Jeremy Weinstein observe, the theories and models described in this chapter are not necessarily contradictory; in fact "different logics of participation may coexist in a single civil war."[92] When considering conflict as relational, or a process involving dynamic interaction between the actors involved, it is likely that different theories and models of mobilization are more applicable at different points in the life cycle of a resistance movement.

General observations suggest that mobilization in the early, middle, and late stages of an insurgent group relies on different strategies according to changing security risks. In the early stages of a movement, leaders seek to carefully select, investigate, and approach potential fellow insurgents. Because security risk is particularly important, leaders begin by mobilizing trusted individuals in their social networks. Afterward, leaders may seek out individuals or groups that share their beliefs. In the middle phases of an insurgency, when security risks are less pronounced, leaders usually have to expand the recruiting effort to meet growing operational and functional requirements and to replace members lost to attrition. Leaders might form coalitions or alliances with other movements that broaden the available networks for recruitment. In the later stages, when an insurgency is a potent challenge to the state, recruiting is characterized by the momentum of the movement. As its security risks decline with increasing territorial control, it can rely more on low-commitment individuals motivated by financial or other selective incentives to rapidly surge the effort needed to prevail against a powerful state opponent.[93]

One method frequently used to break down the life cycle of a resistance movement is breaking it down into different phases or periods where the movement exhibits characteristics peculiar to each phase. Based on a comprehensive review of existing literature on the subject, the ARIS research team developed its own phasing schemata, depicted in Figure 2.

The ARIS phasing construct includes four key phases that a resistance movement can inhabit at various points in its life cycle, including the preliminary, incipient, crisis, and institutionalization phases. It is important to note that while Figure 2 depicts the phasing in a linear progression, a resistance movement can move back and forth between the phases in a nonlinear progression.

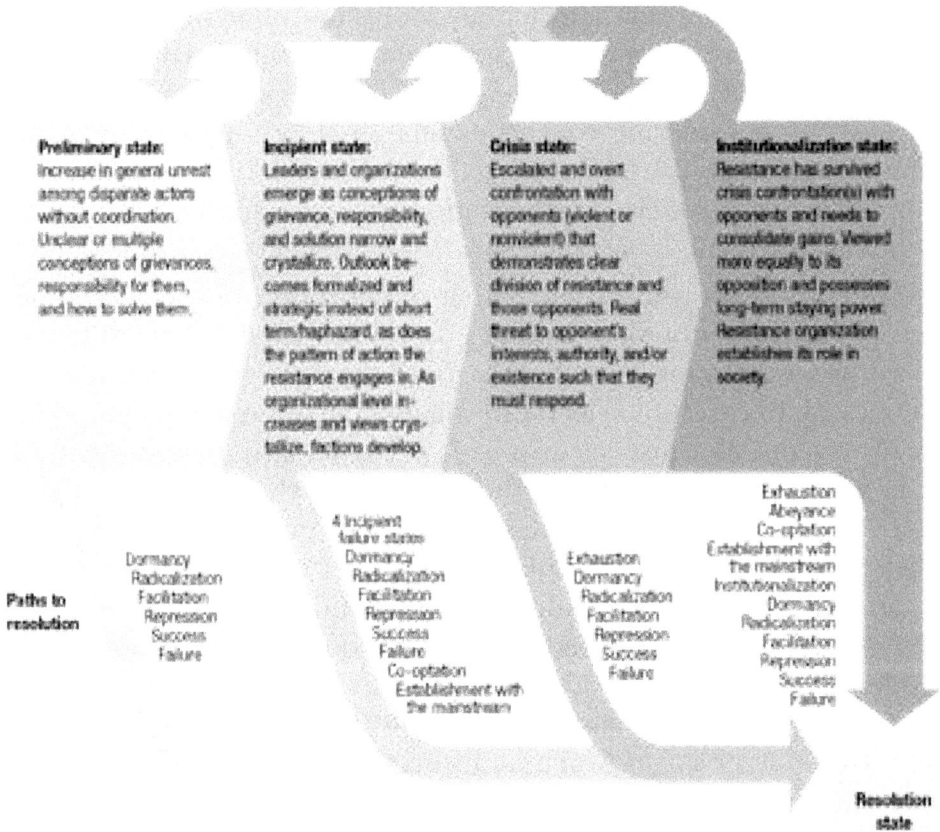

Figure 2. ARIS proposed states for phasing construct analysis.

The limited scope of this work prohibits a detailed discussion of each phase and the mobilization processes likely to occur in each phase. The following discussion on phasing and mobilization is suggestive only, requiring additional research to empirically verify the conclusions reached here. For now, two phases, the incipient and crisis phases, are introduced, and preliminary correlations are drawn between the phase and which mobilization theories and models are likely to be present in it. The incipient phase, the second phase in the ARIS construct, is when mobilization processes become especially important. This phase marks the stage at which the movement first begins to intentionally develop as an organization and adopts a narrative unifying the group. It is when common ideas are first put into collectively based action. In the next phase, the crisis phase, the organization first begins to engage in confrontations with its opponent.

Mobilization in Underground or Clandestine Organizations

Figure 3 depicts a quadrant graph of key features of a state that impact the trajectory of a resistance movement: state strength versus state weakness and state openness versus state closedness.

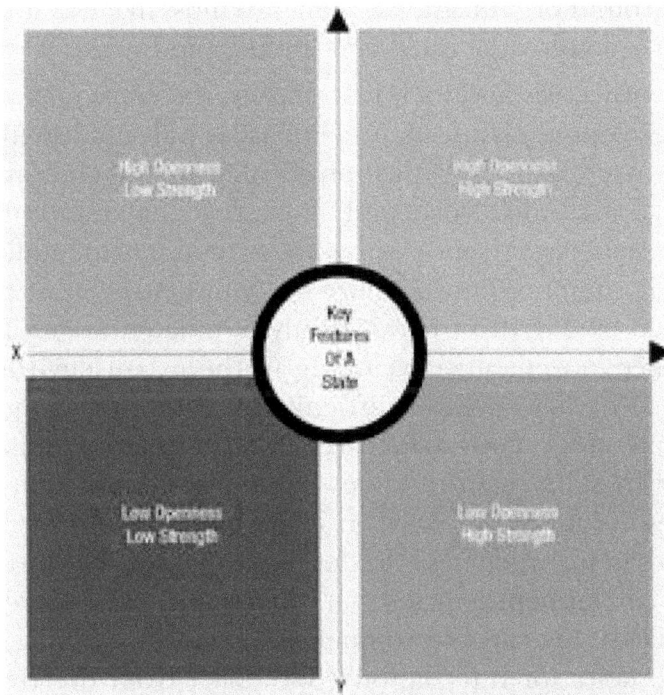

Figure 3. Quadrant graph of key features of a state.

A state's strength reflects its capacity to enforce its policies within its territory, including quelling political opposition. State weakness, on the other hand, signals a state's inability to ensure that its policies and coercive power are projected throughout the state. Instead, the state might enjoy a monopoly on force only in the capital region or other centers of power. A state's level of openness, on the other hand, refers to the extent of a state's policies on tolerating political opposition. In very open states, such as the United States, so long as an opposition group plays by the rules of the game, most forms of political opposition are tolerated. Unless an opposition group chooses to use illegal tactics, such as violence, there is little reason for the opposition group to operate clandestinely. More closed states, however, have little tolerance for political opposition. In these states, participation in such groups is a risky endeavor because individuals risk arrest, imprisonment, or death. Under these circumstances, leaders are far more likely to adopt a clandestine underground movement. However, it is important to note that some states may be closed but also demonstrate characteristics of a weak state, making it unlikely that the security or intelligence forces have the capacity to effectively shut down political opposition. Clandestine underground organizations, then, are most likely to form in states that fall in the lower right quadrant of the graph.

In the early crisis and incipient phases, the resistance movement's focus on developing clandestine elements is particularly likely. When confronting a strong state with sophisticated intelligence and coercive capabilities, a resistance movement has a great deal of incentive to keep its activities and organization secretive or compartmentalized. Moreover, a strong state presence is more likely in urban areas, particularly in the capital city, than in rural, thinly populated areas. This means that resistance movements developing in the urban environs, such as Colombia's M-19, have greater difficulty transitioning to more mature phases that leverage mass armed rebellion because the likelihood of interdiction by state security forces is an ever-looming threat. Urban insurgencies are rare compared to insurgencies based in rural areas. However, urban insurgencies may develop when security forces are constrained by government policies that hinder the use of repressive force, even if they have the capacity to quell the opposition. The government may fear domestic or international blowback from overtly repressive countermeasures.[94] However, in weak states, the need for covertness is less pressing. The insurgent group al-Shabab, for instance, operates in the notoriously weak state of Somalia, which falls in the lower left quadrant of the graph. The inability of the federal government to project its power over even areas within the capital city, Mogadishu, means that the group has little need to operate covertly, particularly in the rural areas.

Given these conditions, preconflict mobilization in the early incipient and crisis phases within clandestine organizations faces peculiar challenges. As O'Connor and Oikonomakis explain, "pre-conflict mobilization . . . occurs in the context of ambiguous constraints and opportunities, which have often become clarified by the time subsequent recruits become involved."[95] It is also a time in which burgeoning resistance movements are vulnerable to repression because of their inexperience, small size, and limited support and logistical capabilities.[96] Indeed, most resistance movements find it difficult to move past these initial phases. Moreover, because of their resource constraints, clandestine organizations cannot rely on the promise of financial gains to their participants. Under these conditions, however, indiviudals are also likely not recruited based solely on their ideological commitments.

The clandestine nature of resistance movements in their early phases impacts recruitment or mobilization processes. Recruitment into the secretive organization necessitates strong feelings of trust, affiliative bonds, and identification between the recruiter and the individual targeted for recruitment when organizations form in strong states with low levels of openness to political opposition. Characteristics of clandestine organizations contribute to these dynamics. Undergrounds operate in an inherently high-risk environment, requiring a high level of group identification to operationally and cognitively mitigate the risk. Viterna, in her study of demobilized fighters from the FMLN in El Salvador, found that existing social networks were important recruiting mechanisms in the early stages of the conflict. In her interviews with former FMLN officials, recruiters reported targeting existing networks deemed sympathetic to the FMLN to ensure that new recruits could be trusted not to inform on existing clandestine supporters to the Salvadoran Armed Forces.[97]

Colombia's M-19: Recruitment in Clandestine Organizations

The Colombian leftist guerilla movement, M-19, remained an urban, clandestine organization for most of its operational history. Many of the early recruits, as well as later recruits, were targeted at Colombian universities. The universities acted as a sort of clandestine job fair for insurgent groups in Colombia, including M-19, the ELN, and the FARC. The university setting offered leftist recruiters ample opportunity to observe, interact with, and ultimately draw in sympathetic individuals. The recruitment process could last several years, and most recruits were unaware that they were being targeted for recruitment into clandestine organizations until late in the process.

M-19 used a common risk-averse approach to recruitment by targeting individuals who were already politically active in legal organizations that had objectives and ideologies similar to those of the clandestine organization. Many of the original M19 leaders belonged to the Juventudes Comunista (JUCO), or Communist Youth Movement, establishing a solid social network among the founding members. With the help of this existing network, the founding members later participated in the FARC, the ELN, and Alianza Nacional Popular, or National Popular Alliance (ANAPO). When recruiting members, M-19 drew heavily from various political and social student organizations that exhibited leftist sympathies. Recruiting from these organizations provided a ready pool of recruits with existing ties to known members and appropriate ideological affinities, both conditions which decreased the likelihood that recruits were covert agents. The danger of infiltration by covert agents was a problem, particularly after several high-profile M-19 operations resulted in the killing of a labor union leader and the theft of thousands of weapons from the Colombian army.

The recruitment process was typically slow. Known operatives carefully watched and vetted potential recruits, sometimes for as long as several years, before formally approaching them with invitations to join the clandestine organization. One former M-19 insurgent reported that she was unwittingly courted by M-19 for nearly two years before receiving a formal invitation. After she accepted the invitation to join, she discovered that the vast majority of her fellow compatriots in her leftist student organization were in fact M-19 members. Before officially being accepted into the organization, recruits were also given assignments to test their mettle and value to the organization.

Summer Newton, ed., *Case Studies in Insurgency and Revolutionary Warfare: Colombia (1964–2009)* (Fort Bragg, NC: USASOC, 2012), 247–250.

Covert resistance in the early phases of a resistance movement also impacts group dynamics in other ways. Clandestine organizations require an unusually high level of commitment from the individual, including a sweeping change in lifestyle that would otherwise be infeasible, or more difficult, in the absence of a strong group identity. This group identification tends to further solidify over time as "the excitement of shared risks strengthens friendships."[98] Clandestine organizations rely on strong ties to increase participation, while movements operating in open political environments may rely on weak ties to form mass-based movements that are more successful in achieving their political objectives. As a result, while the clandestine organizations are compact and exhibit more homogenous interests, motivations, and identities among participants, mass-based movements can evidence more heterogeneous membership.

Recruitment in the early phases of clandestine movements is more likely to occur in safe territories. While the concept of safe territories is more often used to explain the persistence of an insurgency, such as when an armed actor enjoys a safe haven in a weak or welcoming state, safe territories also play a role in recruitment and mobilization processes.[99] Lorenzo Bosi describes a safe territory as a "physical space . . . in which social networks develop over time and shape formal and informal infrastructures of support that maintain dense affective, familial, and personal relations between armed activists and their local constituencies."[100] While mature resistance movements with significant armed components, such as the FARC in Colombia, might have safe territories that extend over significant geographical territory, clandestine organizations generally have much less room to maneuver. In urban areas, a safe territory might be a university campus or nationalist neighborhood enclaves. O'Connor and Oikonomakis describe how the Partiya Karkerên Kurdistan (PKK), the Kurdish Workers Party in Turkey, and the Fuerzas de Liberación Nacional/Ejército Zapatista de Liberación Nacional (FLN/EZLN) in Mexico first mobilized supporters in safe territories provided by local universities. The initial mobilization in both resistance movements began with radical students, such as the PKK's Abdullah calan, who diligently cultivated dense social networks that served as facilitators of recruitment, targeting "preexisting networks of friends and family and from those with a past history of activism." While most were Kurdish, a significant number were also Turkish. The FLN/EZLN, led by César Germán Yáñez Muñoz, recruited in the "safe" university setting where he established a network of trusted recruits.[101]

Mobilization in the early phases is also likely to involve certain types of individuals referred to as early risers or early mobilizers.[102] These individuals are critical to altering emotional and cognitive functions of the targeted audience that can have significant impacts on the possibility of more widespread mobilization in the future. In high-risk mobilization, such as in the early stages of a resistance movement or in clandestine groups, early mobilizers are critical. Early mobilizers or risers have higher thresholds of fear and more robust motivations for participation. Early mobilizers are thought to initiate an emotive process called fear abatement when engaging in initial protest or resistance actions. The actions of the early mobilizers help to abate fear by alerting the more timid, quiescent members of the population that others are dissatisfied with the regime and that fearful individuals would not be alone in targeting the regime:

> Their actions break the reign of preference-falsification that undergirds the regime, and leads to the second step of progressive, larger mobilizations . . . assuming repression is not increased.[103]

Early risers have identities similar to Viterna's strong participant identity. Individuals with a strong participant identity have generally been involved in political activist organizations for some time. The identity is strong because it takes precedence over other identities available to individuals. In her study of female former guerrillas in the FMLN, some women with strong participant identities overcame significant barriers to participation, including identities or roles associated with motherhood. One woman, a mother, reported mobilizing with her husband and her two children, aged seven and nine.[104]

Discussing the authoritarian Franco regime in Spain, Hank Johnston emphasizes the critical role that early risers played, noting that the first protests, highly vulnerable to repression, were attended by individuals with higher thresholds for fear.[105] Though low in number, the early risers understood that people witnessing public discontent was a crucial component of propelling the movement's maturity despite its relatively small size. The early protests had a different purpose than later protests. The first protests were symbolic actions, intended to communicate to the population held in check by fear that resistance is possible and less risky than they perceive it to be. The symbolic protests targeting the Franco regime included flag placements, graffiti writing, unauthorized singing at concerts, and mass shredding of official newspapers. The symbolic actions initiate cat-and-mouse games with the police often looking "foolish and incompetent," while protesters are perceived as daring and confident in comparison.[106]

NONVIOLENCE AND MOBILIZATION

How does mobilization in violent campaigns differ from mobilization in nonviolent campaigns? The question has become of particular importance to social scientists as research suggests that nonviolent campaigns have been more successful than violent campaigns in the past century. Analyzing 323 violent and nonviolent resistance campaigns that occurred between 1900 and 2006, Erica Chenoweth and Maria Stephan found that nonviolent resistance movements were nearly twice as likely as their violent counterparts to achieve all or some of their aims.[107] In particular, nonviolent resistance campaigns were more likely to achieve their objectives when the goal was regime change or liberation from a foreign power, but they were less successful when secession was the objective.[108] Additionally, the authors observed that transitions to democracy were more stable and less likely to culminate in civil war if the transition was nonviolent, as opposed to violent.[109]

What accounts for these differences in outcomes between violent and nonviolent resistance movements? The authors argue that the latter were relatively more successful in terms of mobilizing a broad and diverse segment of the population against a governing authority. The enhanced numbers in turn facilitated a number of processes necessary for successful resistance. Specifically, the authors note that the physical, moral, informational, and commitment-related barriers to participation were lower for nonviolent movements.[110] Regarding the physical requirements to participation, the researchers note that participation "does not require strength, agility or youth. Participation in a nonviolent campaign is open to female and elderly populations, whereas participation in a violent resistance campaign is often, though not always, physically prohibitive."[111]

Nonviolent campaigns also typically have an informational advantage. Compared to violent movements, nonviolent campaigns rely less on underground activities, and therefore nonparticipants are able to observe open and growing collective acts of resistance, which in turn may lead them to conclude that participation is less risky.[112] Similarly, nonviolent campaigns can easily incorporate the participation of those with relatively lower levels of commitment and risk tolerance. Unlike participation in a violent campaign, participants in a nonviolent campaign are much less likely to face death. The relative anonymity of participating in large, nonviolent groups also enables participants to return to their jobs, families, and daily lives with less likelihood of disruption or major sacrifice.[113] Lastly, ethical concerns over tactics are more likely to limit participation in violent movements than in nonviolent ones.[114]

Larger and more diverse participation in turn facilitates various mechanisms that increase the chances of success for a nonviolent movement. First, the more diverse the participation in terms of gender, age, ethnicity, religion, profession, ideology, and socioeconomic status, the more difficult it will be for a governing authority to isolate the movement from the rest of society.[115] Additionally, the more diverse and broad-based a movement, the higher the likelihood that members of the resistance share kinship ties or other connections with armed forces personnel and security and economic elites, who typically represent key pillars of regime support. These links in turn may lead to less elite support for crackdowns on nonviolent opposition and more support for negotiations and concessions to the opposition.[116] Nonviolent movements are more likely to receive international support, while the use of violence against a nonviolent movement is more likely to result in international sanctions against the offending regime. The diversity of participation involving various societal sectors and groups is more likely to generate tactical diversity and innovation, which keeps a governing power off balance.[117]

While Chenoweth and Stephan's research has helped to launch exciting research programs comparing nonviolent and violent campaigns, there are some challenges to the evidence presented by the researchers. Some argue, for instance, that the results of Chenoweth and Stephan's research are unreliable because resistance movements may transition to violence in especially "hard" cases when facing state repression. By contrast, resistance movements choose nonviolent strategies in "easy" cases where brutal regime repression is less likely. As a result, the success of a nonviolent campaign is more reflective of the existing structural conditions than the efficacy of the strategy itself. In his study of the impact of indigenous peasant movements on authoritarian rule in Mexico, Guillermo Trejo found that the movement shifted to violent strategies as the brutality of the government increased.[118]

Violent mobilization is also more likely in countries with fewer resources, whether political opportunities, financial resources, or human resources, while nonviolent mobilization occurs more in resource-rich environments. As a result, violent mobilization emerges in states with underdeveloped countries, low educational attainment, and poor state capacity. Nonviolent mobilization, by contrast, is well suited for states in more urban and developed states that have good prospects for mass mobilization.[119] The external conditions that shape the onset of violent and nonviolent mobilization, including repression and resource availability, can also impact the failure and success of the movements themselves. Furthermore, because violent strategies can exact high costs to governments without the mass mobilization necessary in nonviolent campaigns, resistance leaders might adopt violent strategies to compensate for lower numbers when mass mobilization is less likely.[120]

CONCLUSION

The key resource of any resistance movement is its people. However, mobilizing recruits for participation in resistance movements remains one of the most significant and pressing challenges for resistance leaders. Collective action requires uniting individuals with separate, and sometimes competing, interests, identities, and localities. Uniting a disparate group of people into a cohesive unit capable of acting together requires sophisticated social solutions. Explaining the how and why of those social solutions is among the most intriguing puzzles in the science of resistance. Social scientists have spent decades finding explanations for how some resistance movements are able to overcome barriers to collective action.

There are numerous theories about what makes mobilization an especially difficult endeavor. One of the earliest, and most influential, is Olson's free-rider problem that describes how the creation of public goods disincentivizes individuals from participating in costly resistance activities. Others argue that the key to overcoming the simple logic in Olson's theory is understanding selective incentives. Resistance leaders offering lucrative resources, whether in terms of salary, loot, or natural resources, are able to persuade others to join. The incentives provide recruits with sufficient private gain to overcome the disincentive of the creation of a public good. When the theory of selective incentives explains mobilization, it means that personal greed, not ideology or grievances, is the primary motivator for resistance.

However, the free-rider and selective incentive theories are based on the assumption that people are rational actors pursuing the avenue that reaps the highest benefits, or self-interest, and the lowest cost. Others challenge the assumption that individuals are primarily rational and instead offer more emotive or affiliative explanations for mobilization. Social networks and affiliative factors provide robust explanations for recruitment or mobilization. Individuals are drawn to resistance movements through kinship ties formed in social networks of friends and family. The longing for belonging and acceptance, or affiliative needs, also drives mobilization. Sometimes, affinity with the movement's ideology comes only after the individual has joined. Emotions, or in-process benefits, have also been found to be powerful mobilizers. Some are drawn to participation in resistance movements not because of any expected future payoffs but from emotional satisfaction obtained while participating. Others, however, are prompted to join by emotions of indignation or humiliation.

Some speculate that psychological factors or ideology play a role in facilitating mobilization processes. While it is common for observers to attribute psychological or cognitive disorders to individuals involved in resistance, particularly when indiscriminate or terrorist violence is involved, researchers have not identified any psychological or demographic profiles that predict who will join such movements. While many theories related to mobilization do not explicitly include ideology, ideas can have significant impacts on mobilization processes. Ideology, a comprehensive narrative system that incorporates a society's existing norms, values, and beliefs, is a potent resource for resistance leaders. It can help incorporate locally occurring events into a larger narrative framework that directs attention to the movement's goals and solutions. Moreover, when adapting or distorting societal beliefs, ideology can help justify sanctioned behavior, such as suicide bombings or killing in-group members.

More recently, political scientists have looked at conflicts as dynamic processes that undergo large shifts across their life cycles. When shifts occur, they alter the interests, preferences, and identities of the actors involved, which means that different logics of participation are likely to be prevalent at the same time or in different stages of the conflict. In the control-collaboration model, the dynamics of violence and territorial control influence individual preferences. When territorial control is contested, individuals in affected communities are more likely to mobilize out of fear-based mechanisms. Different resistance phases also trigger different logics of participation. In the ARIS phasing spectrum, the clandestine or underground activities of movements in the incipient or crisis phases are likely to rely on social networks, affiliative factors, and individuals with lower risk perceptions to counteract the group's high security risks and the uncertainty inherent in underground activities.

ENDNOTES

[1] Portions of this chapter are adapted, revised, and updated from Nathan Bos, ed., Human Factors Considerations of Undergrounds in Insurgencies, 2nd ed., chapters 5–7.

[2] Sidney G. Tarrow, *Power in Movement: Social Movements and Contentious Politics* (New York, NY: Cambridge University Press, 2011), 8.

[3] William A. Gamson, *The Strategy of Social Protest* (Homewood, IL: Dorsey Press, 1975), 15.

[4] Ziad W. Munson, *The Making of Pro-life Activists: How Social Movement Mobilization Works* (Chicago, IL: University of Chicago Press, 2010), 4.

[5] J. Bowyer Bell, "Dragonworld (II): Deception, Tradecraft, and the Provisional IRA," *International Journal of Intelligence and Counterintelligence* 8, no. 1 (1995): 21–50.

[6] Resistance movements adopt an important strategy, called identity work, to overcome barriers to participation. Identity work is discussed in chapter 4.

[7] More recently, scholars have begun to discuss the logic of connective action associated with resistance movements, such as those associated with the Arab Spring or the Occupy Wall Street movement, that mobilize through various social media platforms. The distinguishing characteristics of connective action include a lack of formal organizations, simplified master frames, and the absence of a robust collective identity. This distinction and the role that social media plays in contemporary resistance movements in the cyber domain is the topic of a forthcoming edited volume, *ARIS Resistance in the Cyber Domain.*

[8] See W. Lance Bennett and Alexandra Segerburg, *The Logic of Connective Action: Digital Media and the Personalizatin of Contentious Politics* (Cambridge, UK: Cambrige University Press, 2013).

[9] Mancur Olson, *The Logic of Collective Action* (Cambridge, MA: Harvard University Press, 1971), 1.

[10] Olson defines public goods as "any good such that, if any person . . . consumes it, it cannot feasibly be withheld from others in that group."

[11] Olson, *The Logic of Collective Action*, 14.

[12] Stathis N. Kalyvas and Matthew Adam Kocher, "How 'Free' is Free Riding in Civil Wars? Violence, Insurgency, and the Collective Action Problem," *World Politics* 59, no. 2 (2007): 177–216.

[13] Samuel L. Popkin, *The Rational Peasant: The Political Economy of Rural Society in Vietnam* (Berkeley, CA: University of California Press, 1979).

[14] Mark Lichbach, *The Rebel's Dilemma* (Ann Arbor, MI: University of Michigan Press, 1998), 216.

[15] Paul Collier and Anke Hoeffler, "Greed and Grievance in Civil War," *Oxford Economic Papers* 56 (2004): 563–595.

[16] Thank you to Dr. Elizabeth McClintock for this clarification.

[17] Päivi Lujala, Nils Petter Gleditsch, and Elisabeth Gilmore, "A Diamond Curse? Civil War and a Lootable Resource," *Journal of Conflict Resolution* 49, no. 4 (2005): 538–562.

[18] Paivi Lujala, "The Spoils of Nature: Armed Civil Conflict and Rebel Access to Natural Resources," *Journal of Peace Research* 47, no. 1 (2010): 15–28.

[19] Abraham Harold Maslow, "A Theory of Human Motivation," *Psychological Review* 50, no. 4 (1943): 370.

[20] Chuck Crossett and Jason Spitaletta, *Radicalization: Relevant Psychological and Sociological Concepts* (Ft. Meade, MD: Asymmetric Warfare Group, 2010).

[21] Aidan Kirby, "The London Bombers as 'Self-Starters': A Case Study in Indigenous Radicalization and the Emergence of Autonomous Cliques," *Studies in Conflict and Terrorism* 30, no. 5 (2007): 415–428.

[22] One researcher noted that activists in the pro-life movement, a movement associated with especially fervent beliefs, often become involved before they prescribe to the movement's beliefs about abortion; some were even pro-choice when they first joined. The beliefs and ideology of the activists, however, were altered to include pro-life beliefs about abortion as a result of their participation in the movement.

[23] Munson, *The Making of Pro-life Activists*,

[24] Marc Sageman, *Understanding Terror Networks* (Philadelphia, PA: University of Pennsylvania Press, 2004).

[25] The Maghreb is a term referencing the region of western North Africa. The researcher distinguishes between the "Core Arab" region and "the Maghreb Arab" region

[26] Reinoud Leenders, "Collective Action and Mobilization in Dar'a: An Anatomy of the Onset of Syria's Popular Uprising," *Mobilization: An International Journal* 17, no. 4 (2012): 419–434.

[27] Elisabeth Jean Wood, *Insurgent Collective Action and Civil War in El Salvador* (Cambridge: Cambridge University Press, 2003), 18.

[28] Ibid., 18.

[29] William A. Gamson, "Constructing Social Protest," in *Social Movements and Culture*, ed. Hank Johnston and Bert Klandermas (Minneapolis: University of Minnesota Press, 1995), 89.

[30] Jeff Goodwin, James M. Jasper, and Francesca Polletta, "Emotional Dimensions of Social Movements," in *The Blackwell Companion to Social Movements*, eds. David A. Snow, Sarah A. Soule, and Hanspeter Kriesi (Malden, MA: Blackwell Publishing, 2007), 422.

[31] James M. Jasper, *The Art of Moral Protest* (Chicago: University of Chicago Press, 1997), 140.

[32] James M. Jasper and Jane Poulsen, "Recruiting Strangers and Friends: Moral Shocks and Social Networks in Animal Rights Activism and Anti-Nuclear Protests," *Social Problems* 42, no. 4 (1995): 493–512; Elisabeth Jean Wood, *Insurgent Collective, Action and Civil War in El Salvador* (Cambridge, UK: Cambridge University Press, 2003), 13–15.

[33] Jeff Goodwin and Steven Paff, "Emotion Work in High-Risk Social Movements," in *Passionate Politics: Emotions and Social Movements*, eds. Jeff Goodwin, James M. Jasper, and Francesca Polletta (Chicago: University of Chicago Press), 286–287.

[34] Goodwin et al., "Emotional Dimensions of Social Movements," 416–417.

[35] Rona M. Fields, *Martyrdom: The Psychology, Theology, and Politics of Self-Sacrifice* (New York: Praeger, 2004).

[36] Jeff Victoroff and Arie Kruglanski, *Psychology of Terrorism: Classic and Contemporary Insights* (Washington, DC: Psychology Press, 2009).

[37] Sverre Varvin, "Humiliation and the Victim Identity in Conditions of Political and Violent Conflict," *Scandinavian Psychoanalytic Review* 28, no. 1 (2005): 40–49.

[38] Richard E. Nisbett and Dov Cohen, *Culture of Honor: The Psychology of Violence in the South* (Boulder, CO: Westview Press, 1996).

[39] Ibid.

[40] Grosjean, Pauline Grosjean, "A History Of Violence: The Culture of Honor and Homicide in the US South," *Journal of the European Economic Association* 12, no. 5 (2014): 1285–1316; P. J. Henry, "Low-status Compensation: A Theory for Understanding the Role of Status in Cultures of Honor," *Journal of Personality and Social Psychology* 97, no. 3 (2009): 451.

[41] Clark R. McCauley and Sophia Moskalenko, *Friction: How Radicalization Happens to Them and Us* (New York: Oxford University Press, 2011).

[42] Henry, "Low-status Compensation," 451.

[43] Roxanne L. Euben, "Humiliation and the Political Mobilization of Masculinity," *Political Theory* 43, no. 4 (2015): 1–33.

[44] The ummah is an Islamic term for the whole community of Muslim believers.

[45] Euben, "Humiliation and the Political Mobilization of Masculinity," 1–33.

[46] Leenders, "Collective Action and the Mobilization in Dar'a."

[47] Alexis Leanna Henshaw, "Where Women Rebel: Patterns of Women's Participation in Armed Rebel Groups 1990–2008," *International Feminist Journal of Politics* 18, no. 1 (2016): 39–60.

[48] Reed M. Wood and Jakana L. Thomas, "Women on the Frontline: Rebel Group Ideology and Women's Participation in Violent Rebellion." *Journal of Peace Research* 54, no. 1 (2017): 31–46.

[49] Jocelyn Viterna, "Pushed, Pulled, and Persuaded: Explaining Women's Mobilization Into the Salvadorian Guerilla Army," *American Journal of Sociology* 112, no. 1 (2006).

[50] Jocely Viterna, "Radical or Righteous? Using Gender to Shape Public Perceptions of Political Violence," in *Dynamics of Political Violence: A Process-Oriented Perspective on Radicalization and the Escalation of Political Conflict*, eds. Lorenzo Bosi, Chares Demetriou, and Stefan Malthaner (London and New York: Routledge, Taylor & Francis Group, 2014), 190–191.

[51] Henshaw, "Where Women Rebel," 204–219.

[52] Portions of this section are adapted, revised, and updated from Nathan Bos, ed., Human Factors Considerations of Undergrounds in Insurgencies, 2nd ed., chapter 7.

[53] Walter Lacquer, *A History of Terrorism*, (New Brunswick: Transaction Publishers, 2012), 30.

[54] Chuck Crossett and Jason A. Spitaletta, *Radicalization: Relevant Psychological and Sociological Concepts* (Laurel, MD: Johns Hopkins Applied Physics Laboratory).

[55] Ibid.

[56] Andrew Silke, "Cheshire-cat Logic: The Recurring Theme of Terrorist Abnormality in Psychological Research," *Psychology, Crime and Law* 4, no. 1 (1998): 51–69.

[57] Georges Sorel, *Reflections on Violence* (Cambridge: Cambridge University Press, 1999), 14.

[58] Portions of this section are adapted, revised, and updated from Nathan Bos, ed., *Human Factors Considerations of Undergrounds in Insurgencies*, 2nd ed., chapter 5.

[59] Sara Savage and Jose Liht, "Mapping Fundamentalisms: The Psychology of Religion as a Sub-discipline in the Understanding of Religiously Motivated Violence," *Archive for the*

Psychology of Religion 30, no. 1 (2008): 75–91; Angela J. Yu and Peter Dayan, "Uncertainty, Neuromodulation, and Attention," *Neuranatomy* 46, no. 4 (2005): 681–692.

[60] Jason Spitaletta, "Egyptian Islamic Jihad." in Chuck Crossett (ed.), *Casebook on Insurgency and Revolutionary Warfare, Volume II: 1962–2009* (Fort Bragg, NC: USASOC, 2012).

[61] Ibid.

[62] This study was conducted prior to the rise of ISIS, which constitutes the ongoing phase of global Salafist jihad discussed in this text box case study.

[63] Johannes J. G. Jansen, *The Neglected Duty: The Creed of Sadat's Assassins and Islamic Resurgence in the Middle East* (New York: Macmillan, 1986).

[64] Devin R. Springer, L. Regens, and David N. Edger, *Islamic Radicalism and Global Jihad* (Washington, DC: Georgetown University Press, 2009).

[65] Fawaz A. Gerges, *The Far Enemy: Why Jihad Went Global* (New York: Cambridge University Press, 2005).

[66] Spitaletta, "Egyptian Islamic Jihad."

[67] Ibid.

[68] Jason Spitaletta and Shana Marshall, "Al Qaeda," in Chuck Crossett (ed.), *Casebook on Insurgency and Revolutionary Warfare, Volume II: 1962–2009* (Fort Bragg, NC: USASOC, 2012).

[69] Muntasir Zayyat, *The Road to Al-Qaeda: The Story of Bin Laden's Right-Hand Man* (London: Pluto Press, 2004).

[70] Spitaletta and Marshall, "Al Qaeda."

[71] Yuichi Kubota, *Armed Groups in Cambodian Civil War: The Stronghold and Beyond* (New York: Palgrave Macmillan 2013), 1.

[72] Stathis Kalyvas and Matthew Adam Kocher, "How 'Free' is Free Riding in Civil Wars? Violence, Insurgency, and the Collective Action Problem," *World Politics* 59, no. 2 (2007): 213.

[73] Stathis N. Kalyvas and Matthew Adam Kocher, "The Dynamics of Violence in Vietnam: An Analysis of the Hamlet Evaluation System (HES)," *Journal of Peace Research* 46, no. 3 (2009): 335–355; Gonzalez Vargas, "Urban Irregular Warfare and Violence Against Civilians: Evidence From a Colombian City," *Terrorism and Political Violence* 21, no. 1 (2009): 110–132.

[74] Stathis N. Kalyvas, "Micro-Level Studies of Violence in Civil War: Refining and Extending the Control-Collaboration Model," *Terrorism and Political Violence* 24, no. 4 (2012): 658–668.

[75] Ibid.

[76] Ana M. Arjona and Stathis N. Kalyvas, "Recruitment into Armed Groups in Colombia: A Survey of Demobilized Fighters," in *Understanding Collective Political Violence* ed. *Yvan Guichaoua* (London: Palgrave McMillan, 2011), 167.

[77] Kalyvas and Kocher, "How 'Free' is Free Riding in Civil Wars."

[78] Ibid.

[79] Ibid.

[80] This is not to say, however, that internally displaced persons and refugees do not also incur high costs in fleeing violence.

81 The term *irregular warfare* as it is used here is just meant to indicate warfare in which one or more nonstate actors are engaged in a conflict with another actor, generally a state actor.

82 Stathis Kalyvas, *The Logic of Violence in Civil War* (New York: Cambridge University Press, 2006), 89–91.

83 Kalyvas and Kocher, "How 'Free' is Free Riding in Civil Wars."

84 David T. Mason and Dale A. Krane, "The Political Economy of Death Squads: Toward a Theory of the Impact of State-sanctioned Terror," *International Studies Quarterly* 33, no. 2 (1989): 175–198.

85 As quoted in Viterna, "Pulled, Pushed, and Persuaded," 1–45.

86 Ibid.

87 David T. Mason and Dale A. Krane, "The Political Economy of Death Squads: Toward a Theory of the Impact of State-sanctioned Terror," *International Studies Quarterly* 33, no. 2 (1989): 175–198.

88 David Stoll, *Between Two Armies: In the Ixil Towns of Guatemala* (New York: Columbia University Press, 1993), 20; Timothy Wickham-Crowley, *Guerrillas and Revolution in Latin America: A Comparative Study of Insurgents and Regimes since 1956*, (Princeton, NJ: Princeton University Press, 1992).

89 Stoll, *Between Two Armies*, 20.

90 For an extended illustrative example of the control-collaboration model in the Greek Civil War, see chapter 4.

91 Kalyvas, "Micro-Level Studies of Violence in Civil War."

92 Macartan Humphreys and Jeremy M. Weinstein, "Who Fights? The Determinants of Participation in Civil War," *American Journal of Political Science* 52, no. 2 (2008): 436–455.

93 See Robert Leonhard, ed., *ARIS Undergrounds in Insurgent, Revolutionary, and Resistance Warfare* (Fort Bragg, NC: United States Special Operations Command, 2013), 19–42, for an in-depth analysis of underground recruitment processes.

94 Paul Staniland, "Cities on Fire: Social Mobilization, State Policy, and Urban Insurgency," *Comparative Political Studies* 43, no. 12 (2010): 1623–1649.

95 Frances Patrick O'Connor and Leonidas Oikonomakis, "Preconflict Mobilization Strategies and Urban-Rural Transition: The Case of the PKK and the FLN/EZLN," *Mobilization: An International Quarterly* 20, no. 3(2015): 379–399.

96 Frances Patrick O'Connor, "Armed Social Movements and Insurgency: The PKK and its Communities of Support" (PhD diss., European University Institute, 2014). http://hdl.handle.net/1814/34582.

97 Viterna, "Pulled, Pushed, and Persuaded," 1–45.

98 Donatella della Porta, *Social Movements, Political Violence, and the State: A Comparative Analysis of Italy and Germany* (Cambridge: Cambridge University Press, 1995), 184.

99 O'Connor and Oikonomakis, "Preconflict Mobilization Strategies."

100 Lorenzo Bosi, "Safe Territories and Violent Political Organizations," *Nationalism and Ethnic Politics* 19, no. 1 (2013): 80–101.

101 O'Connor and Oikonomakis, "Preconflict Mobilization Strategies."

102 For an illustrative example of early risers, please see the example in this chapter on the rise of a resistance movement in the Syrian region of Da'ra against the Assad regime in the beginning stages of the Arab Spring in Syria.

[103] Hank Johnston, "The Mechanism of Emotion in Violent Protest," in *The Dynamics of Political Violence: A Process-Oriented Perspective on Radicalization and the Escalation of Political Conflict*, eds. Charles Demetriou, Stefan Malthaner, and Lorenzo Bosi (Burlington, VT: Ashgate, 2014), 27–50.

[104] Viterna, "Pulled, Pushed, and Persuaded," 1–45.

[105] In this study, the individuals' higher threshold of fear is indicated by their participation in these early protests because the majority of the population experienced preference falsification, where they hide dissatisfaction with the incumbent regime for fear of reprisal by security forces.

[106] Hank Johnston, "The Mechanism of Emotion in Violent Protest," in *The Dynamics of Political Violence: A Process-Oriented Perspective on Radicalization and the Escalation of Political Conflict*, eds. Charles Demetriou, Stefan Malthaner, and Lorenzo Bosi (Burlington, VT: Ashgate, 2014), 37–39.

[107] Erica Chenoweth and Maria J. Stephan, *Why Civil Resistance Works: The Strategic Logic of Nonviolent Conflict* (New York: Columbia University Press), 7.

[108] Ibid., 7, 10.

[109] Ibid., 10.

[110] Ibid.

[111] Ibid., 35.

[112] Ibid., 36.

[113] Ibid., 37–38

[114] Ibid., 37.

[115] Ibid., 40.

[116] Ibid., 46–47.

[117] Ibid., 52–56.

[118] Guillermo Trejo, *Popular Movements in Autocracies: Religion, Repression, and Indigenous Collective Action in Mexico* (New York: Cambridge University Press, 2012).

[119] Peter B. White, Dragana Vidovic, Belén González, Kristian Skrede Gleditsch, and David E. Cunningham, "Nonviolence as a Weapon of the Resourceful: From Claims to Tactics in Mobilization," Mobilization: An International Quarterly 20, no. 4 (2015): 471-491.

[120] Ibid.

CHAPTER 4.
WHY ARE SOME RESISTANCE STRATEGIES AND TACTICS USED OVER OTHERS?

Resistance movements face numerous challenges and barriers to achieving their political goals. As discussed in previous chapters, some political conditions hinder or facilitate movements' efforts. Resistance movements operating in authoritarian regimes with little or no political freedoms are more constrained than groups operating in more democratic regimes. However, it is not only the external environment that shapes the future trajectory of a movement; leaders' internal decisions regarding the group's "vision, direction, guidance, coordination, and organizational coherence" also play a role.[1] These decisions, while having an obvious impact on the operation of a movement, also affect a group's decision-making on preferable strategies and tactics.

Leaders make decisions about a wide variety of issues, including internally and externally related matters. They need to make basic decisions about how to structure the group, how to properly resource it, and how to find safe spaces from which to effectively plan and operate. Moreover, all groups operate within a larger host population. Interactions with the host population have repercussions for a group's legitimacy and levels of popular support. Among the most important strategic decisions leaders make is determining if, and when, the group transitions from nonviolent to violent tactics.

Scholars explain most of these decisions, particularly those related to the transition to violence, as resulting from rational calculations. Academic investigations of similar issues, however, reveal how connected seemingly prosaic decisions are to important strategic and tactical decisions regarding the use of violence and its relationship with the host population. This section explores research related to transitions from nonviolence to violence and also discusses how organizational structure, resourcing, and territorial control impact a group's use of violence, the lethality, target, and scale of violence, and interactions with the host population.

The initial ARIS project, *Casebook on Insurgency and Revolutionary Warfare, Volume II: 1962–2009*, contains twenty-three case studies of resistance movements ranging from nonviolent protest during the Orange Revolution in Ukraine to the 1994 Rwandan genocide and the classic leftist insurgent groups of Colombia. The methodology behind the case selection reflects the intention to represent a set of case studies diverse, both geographically and temporally; much changed since the initial Special Operations Research Office (SORO) studies that focused on underground resistance movements from the Second World War and Maoist-inspired communist insurgencies.

In addition to offering this rich array of cases with varying geographies and places in history, the ARIS team expanded the case studies to include different resistance strategies. Resistance occurs in insurgencies, when nonstate armed groups and incumbent governments resort to widespread violence in support of their political goals, but resistance also occurs when thousands of protesters use nonviolent resistance tactics to overturn the results of corrupt runoff presidential elections, as occurred in Ukraine. In other words, the case selection assumed that resistance and violence are not synonymous.

Since that time, research within the social sciences has come to similar conclusions. Whereas a great deal of research related to resistance focused almost exclusively on full-blown civil wars, some scholars now recognize that while civil war is one manifestation of resistance, a long, unfolding process usually preceded the mass violence characteristic civil war.[2] Sometimes, the process was relatively short, measured in years, but in other cases, such as the protracted insurgency dogging Colombia, the processes are found decades in the past during previous episodes of violence. Regardless, the violence observed in civil war or insurgencies oftentimes first began as nonviolent resistance, such as the mass protests in Syria during the Arab Spring that slowly transitioned to a deadly civil war over the course of several years.

While the research on transitions from nonviolence to violence is a relatively small subfield in political science, one promising approach is found in the theory of relational dynamics. The theory differs from the broad, macrostructural factors attributed to the onset of violence in chapter 2 because it is concerned with explaining why violence erupts at certain times and places. It incorporates components of structure by considering how the larger political environment shapes conflict processes, but it also incorporates the agency of the actors party to the conflict. Similarly, relational dynamics also differ from theories that treat violence as the outcome of rational calculations made by movement leaders without reference to how repeated interactions between the resistance movement, competing movements, the state, nonstate actors, and foreign governments shape decision-making.[3]

Relational dynamics assess how repeated interactions between actors involved in conflict processes can lead to the adoption of violence. Interactions occur in numerous arenas, including between resistance movements and the political environment, among movement actors, between movement activists and security forces, and between a movement and a countermovement. Interactions in these domains between relevant actors during intense periods of mobilization, called protest cycles, are especially important. It is during the protest cycle that the mechanisms contributing to the outbreak of violence typically emerge. The mechanisms driving radicalization, including competitive escalation and political outbidding, involve different forms of competition. While it is expected that resistance actors compete with the targeted regime, relational dynamics bring attention to how competition between actors within a movement, and between countermovements, is a surprising driver of radicalization as well.

Once a resistance movement adopts violence as a tactic, the ways in which it manifests are manifold. Violence varies according to a number of different factors, including lethality, target, and scale. In the case of rioting or attacks on property, violence is nonlethal. When violence is lethal, it varies in terms of scale. Some conflicts are far bloodier than others, killing tens of thousands, such as the Bosnian War, while other conflicts result in relatively fewer casualties, such as the several thousand who died in the long-running Irish Republican insurgency in Northern Ireland. Moreover, the targets of armed actors also vary. In some cases, armed actors target civilians at a higher rate than other organizations. Targeting of civilians can result from indiscriminate violence that targets based on suspected categories of people or selective violence that targets individuals based on personal culpability in some sanctioned behavior.

The lethality, type, and scale of violence adopted by insurgent groups is influenced by decisions on organizational structure, resourcing, and territorial control. Leaders generally make decisions on organizational structure according to the specific strategies and goals of the group. While initial decisions relating to organizational structure are related to strategy, organizational structures also impact future strategies and tactics, particularly regarding the use of violence. Hierarchical structures are associated with higher levels of sustained violence. Groups with functionally based structures commit more lethal attacks than regionally based groups. Flat, networked structures, meanwhile, generate more intragroup competition that can be detrimental to the coherence of the group. Violent nonstate groups are often required to organize to maximize secrecy in compartmentalized cells. The demands for secrecy generate the so-called terrorist dilemma in which leaders sacrifice command and control of the group.[4] This inhibits the ability of the leadership to strategically guide the group, potentially leading to greater use of indiscriminate violence that inhibits positive relations with the host population.

Another relevant strategic decision regards the acquisition of safe spaces needed for planning and operating. Some armed groups remain underground, or clandestine, organizations that gain little in terms of territory. Groups successful in gaining territorial control, however, have a greatly expanded operational range. Territorial control provides greater latitude in how groups interact with the host population. In the control-collaboration model, the extent of territorial control is also correlated with a group's reliance on selective, rather than indiscriminate, violence to consolidate gains against its opponents. Territorial control gives armed groups greater access to intelligence and enhanced incentives for civilians to denounce defectors.

In addition to safe spaces, armed groups also require sufficient resourcing to maintain their planning and operational capabilities. While beds, beans, and bullets are the most obvious resources, they also need to pay off corrupt officials or finance shadow governance activities. More than just a means to an end, a group's method of financing impacts its use of violence. Groups with economic endowments, such as lootable resources, are more prone to using indiscriminate violence, targeting civilians, and committing wartime atrocities. By contrast, groups that rely on social endowments use more selective violence and are more likely to be disciplined by norms outlining appropriate and inappropriate forms of combat.

TRANSITIONS FROM NONVIOLENCE TO VIOLENCE

This section discusses how scholars are using the contentious politics research program to advance the study of violent political conflict. Scholars in this field do not regard organized violence as an inherent facet of resistance, but instead assume that the use of violence is one choice among many possible ones. As a result, when the leadership of a resistance movement opts to use violence, it is a choice that requires explanation. In turn, this means that the study of a resistance movement does not begin when the shooting starts but much earlier when the movement first coalesces as an organization concerned with making political claims against its targets. It is not uncommon for full-blown civil wars and insurgencies to emerge from nonviolent protest waves as the violent conflicts in Syria, Libya, Northern Ireland, Kashmir, and Palestine illustrate. However, resistance movements may continue to rely on nonviolent strategies with both armed and political wings.

Nonviolent resistance has long been a method of human conflict. Although the processes of nonviolent resistance have sometimes conincided with particular philosophical and religious considerations, the tactics remain much the same, whether practiced by groups because of their ethical convictions or as a strategic expedient. Nonviolent resistance movements have had numerous successes over the last forty years, including the Solidarity movement in Poland during the 1980s, which organized massive labor strikes that ultimately unseated the ruling communist regime; the 2004–2005 Orange Revolution in the Ukraine, which saw mass mobilization and broad civil disobedience overthrow the results of a corrupt electoral process; and the January 2011 Lotus Revolution in Egypt, when massive protests in Tahrir Square forced President Hosni Mubarak to relinquish his presidency after ruling for thirty years.

Gene Sharp, a prominent scholar of nonviolent movements, defined nonviolent resistance as "a technique of socio-political action for applying power in a conflict without the use of violence."[5] The techniques are outside the boundaries of conventional political processes, such as voting, lobbying, and interest-group organizing. The persuasiveness of nonviolent campaigns derives from the continual, tactical innovation that produces societal disruption. The tactics are a variety of social, political, and economic methods, including boycotts, strikes, protests, sit-ins, stay-aways, and other forms of noncooperation and civil disobedience intended to put pressure on a ruling authority.[6, 7] In the twenty-first century, resistance movements, including those that use nonviolence, have benefitted from global telecommunication networks that link existing social networks, generating novel, virtual social networks in the information domain.[8]

Initially, the research in contentious politics focused almost exclusively on nonviolent resistance movements. Researchers evidenced a selection bias that favored the study of movements attractive to Western liberalism, such as campaigns centered on human rights or gender equality.[9] More recently, the range of empirical research expanded to include violent resistance movements, demonstrating the usefulness applying the findings on nonviolent resistance movements to those with less attractive goals, including campaigns that rely on sustained, organized violence.[10] Drawing on social movement theory, the program offers a rich set of conceptual tools, processes, and mechanisms to better explain how the dynamics of resistance contribute to the adoption of violent tactics.

The rational choice explanations for the transition to violence are helpfully illustrated by the concept of the threshold of violence, as depicted in the equivalent response model in Figure 4.[11]

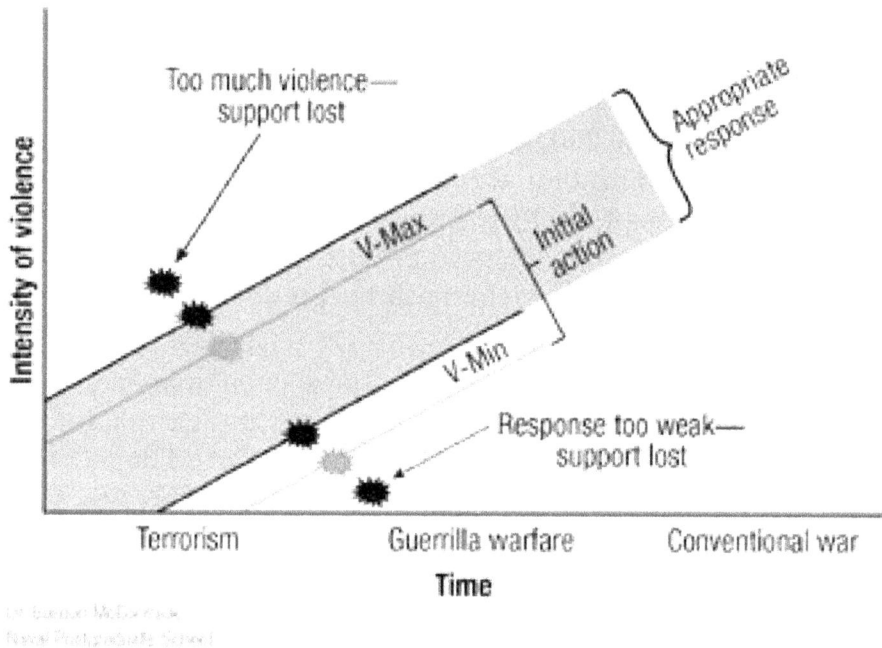

Figure 4. Equivalent response model.

A group's decisions to use violence and how much violence to apply to its target are based on calculations of the relative effectiveness of violence versus other tactics. As shown in Figure 4, resistance movement leaders also need to consider the threshold of violence, or the point at which violent tactics become counterproductive as popular support is lost or too weak to achieve their objectives. In the initial stages of the transition to violence, resistance movements generally begin with smaller scale operations, such as terrorist actions. As the group gains momentum and resources, it may adopt more sophisticated tactics such as small-scale guerrilla actions, culminating in full-blown conventional tactics. This phasing describes the trajectory of the Liberation Tigers of Tamil Eelam (LTTE), which began using violent, irregular tactics in the late 1970s but developed capabilities for conventional tactics by the 1990s.[12, 13] Leaders seek to operate in the band of excellence that balances the need for effectiveness without compromising popular support. However, it is important to note that the state security forces also need to find this band of excellence to effectively counter resistance movement operations without comprising its legitimacy.

Resistance movements rely on various strategies to push that upper threshold of violence to achieve maximum effect in pursuit of their objective. One such strategy is using violence to provoke an extreme government countermeasure that targets population indiscriminately. When the strategy is successful, it generates feelings of injustice and anger among the targeted community, increasing its level of tolerance for insurgent violence. Another common nonviolent tactic is the provision of social services to gain popular support for the group's operations. Lastly, resistance movements can also use narrative framing to expand the community's norms on legitimate targets of violence and acceptable thresholds of violence.

Contentious politics aims to explain organizational transitions from nonviolent to violent tactics and strategies as part of the larger process of radicalization.[14, 15] Explanations within the research emphasize relational dynamics among actors and stakeholders, inside and outside the organization, as well as in the surrounding environment. The process of radicalization unfolds in repeated interactions between actors involved in the conflict. In this regard, radicalization is not seen as emerging from a particular class of people or even a particular ideology. Instead, radicalized groups most often emerge as splinter groups from broader opposition movements. The relational dynamics theory differs from previous theories that treated violence as a rational strategic choice made by movement leaders who maintain violence as the most effective means of achieving their goals. The theory also departs from the macrostructural causes of violence, such as political marginalization and poor socioeconomic development, as the primary precursors to political violence.[16] As Gianluca De Fazio explains:

> [Political] violence does not necessarily emerge from individuals' dispositions or grievances, nor is it the necessary outcome of structural or cultural factors. While these factors shape the context of contention, they cannot explain the timing of violent radicalization.[17]

Radicalization is context sensitive and emerges from interactions among individuals, groups, and institutional actors in numerous arenas of interaction.[18] There are four basic arenas of interaction that are especially important to the process of radicalization: (1) between resistance movements and the political environment, (2) among movement actors, (3) between movement activists and security forces, and (4) between a movement and a countermovement. Some of these arenas of interaction have been treated elsewhere in the text. The first arena, between resistance movements and the political environment, is discussed as a political opportunity structure in chapter 2. This section focuses on the arenas of interaction that have not been fully discussed elsewhere. Interactions within the arenas occur during ongoing protest cycles unleashing causal mechanisms that can contribute to shifts in violence.[19]

Protest Cycle

The radicalization of a previously nonviolent resistance movement occurs within the context of a protest cycle. A protest cycle is a period of heightened or intense mobilization. Sidney Tarrow further defines a protest cycle as:

> a phase of heightened conflict across the social system: with a rapid diffusion of collective action from more mobilized to less mobilized sectors; a rapid pace of innovation in the forms of contention; the creation or new or transformed collective action frames; a combination of organized and unorganized participation; and sequences of intensified information flow and interaction between challengers and authorities.[20]

In the twenty-first century, the most influential protest cycle occurred during the Arab Spring. Protest against the entrenched authoritarian regimes in the Middle East diffused from its origin in Tunisia to less mobilized societies in Libya, Egypt, Syria, and elsewhere in the region.

As the protests spread to other countries, so too did the tactics and techniques that were successful in ousting Tunisian President Ben Ali from power. The raucous street protests in Tunisia shared similarities with those that later brought down President Hosni Mubarak in Egypt.[21, 22] The nonviolent resistance characterizing the early Arab Spring, however, gave way to civil war in Syria as the interactions between protesters and the Syrian security forces became increasingly violent.

A protest cycle is a period of heightened or intense mobilization.

Protest cycles have common characteristics or stages that help to elucidate the process whereby a nonviolent resistance movement becomes radicalized and adopts violence. The first is expansion through diffusion, which is previously described in the example of the Arab Spring. In this stage, early risers are critical in encouraging more mass-based participation by revealing the vulnerabilities of the authorities. As the protest set off by diffusion expands, countermovements emerge that attempt to bolster or restore the status quo. Elites also begin to mobilize, usually deploying a mixed bag of repression and reform policies to counter challenges to their authority. At this juncture in the protest cycle, mass protest may start to slow down, but components of the resistance movement may also radicalize as internal and external pressures brought on by interactions with state authorities and countermovements alter the calculus of factions within the movement.[23] Several mechanisms emerging from arenas of interaction prod the transition from nonviolence to violence, including outbidding, the different pace of demobilization, competitions for power among and between movements, and the formation of countermovements. Before discussing the mechanisms, however, it is necessary to analyze another concept from contentious politics that further illuminates the issue—repertoires of contention.

Repertoires of Contention

Repertoires of contention comprise the cluster of acts resistance movements use to make claims against their targets, usually state actors. The concept is an important one because it helps researchers think about why resistance movements adopt some actions over others, including violence or particular types of violence. Charles Tilly, who first developed the concept, defines it simply as "the ways that people

119

work together in pursuit of shared interests."[24] Each society, culture, historical movement, or other differentiated group has access to a set of stock contentious acts that are familiar and meaningful to the target audience. The acts, because they are culturally embedded within a society, are familiar and resonate with the movement's audience. In Ireland, the hunger strike is a historically and culturally familiar way to protest against injustice, dating back centuries to the Brehon legal system.[25] The Irish Republicans frequently used the hunger strike in prison as a form of political protest against the British, sometimes to great strategic effect.[26, 27] As a concept, repertoires of contention are consciously described in theatrical language to capture the extent to which contentious politics is culturally scripted but also offers room for improvisation in, for example, "loosely scripted theater."[28]

Repertoires of contention are the acts, often clustered, resistance movements use to make claims against their targets.

Successful resistance movements continually seek to creatively adapt existing repertoires to gain tactical advantage, but most adaption generally occurs within already well-established routines within repertoires learned over generations. Despite the long history of hunger strikes in Ireland, other forms of contention are more fleeting and evolve alongside historical conditions. As a result, repertoires differ according to place and time and across groups. The most familiar forms of nonviolent contention, the political march, demonstration, and mass participation, first came into use in nineteenth-century Europe.[29, 30, 31]

Resistance movements deploy three main types of repertoires: violent, conventional, and disruptive. The first, violence, is among the oldest known repertoires.[32, 33] Although in traditional military and defense education, violence is generally depicted as an extreme form of coercion, viewing violence through the lens of repertoires reveals the performative perspective of violence. Not every act of violence perpetrated by a resistance movement is intended to overwhelm a target's security forces. Instead, violence is also deployed to "weld supporters together, dehumanize opponents, and demonstrate a movement's prowess."[34] The perpetration of violence is also used to cultivate a collective identity and a sense of solidarity among participants.[35]

Another frequently used form of contention is disruption. Disruption is a threat of the use of violence. Similar to violence, it has a performative aspect as it signals to others the determination of the movement.

Disruption seeks to interrupt everyday routines of opponents, authorities, and onlookers so that public life cannot seamlessly function as it does under normal circumstances. By disrupting business as usual, whether through blocking traffic or staging sit-ins in public spaces, a resistance movement forces interaction with authorities. Disruption is a powerful but difficult form of contention to sustain. It requires constant innovation on the part of a resistance movement as police or military soon develop tactics to counter it. Disruption is also significant insomuch as it is tactically close to violence. Because disruption requires a high level of commitment, while the moderate rank and file of a movement soon leave to return to their private lives, more militant members remain, who, in turn, are more likely to adopt violent tactics.[36]

In contrast to disruptive forms of contention, conventional forms of contention are widely accepted and used. Some methods, such as strikes and demonstrations, were once disruptive but migrated to conventional repertoires once they were widely institutionalized through legalization in the West. As an example, police in the United States give demonstrators helpful hints on the most effective ways to stage a demonstration. Strikes and demonstrations have become such a familiar part of the contentious landscape that they have acquired a habitualized aspect, from the signs people carry to the slogans and chants they shout. Conventional forms of contention, however, are beneficial because they are able to attract a wide variety of individuals who perceive little danger or uncertainty in participation.[37]

Most repertoires of contention change glacially, shifting as historical epochs draw to a close and new ones open. The advent of print, the emergence of the modern state, and industrialization profoundly impacted how organized challengers protested against political authorities.[38] However, the protest cycle also unleashes numerous mechanisms that lead resistance movements to shift their repertoires of contention to include new or adapted forms of protest. One such shift, of course, is when a movement relying predominantly on nonviolent tactics shifts to predominantly violent tactics. In contentious politics, this process is often referred to as radicalization.

Protest, Repression, and Tactical Innovation

Contentious politics researchers observe that violence emerges as a by-product of repeated interactions between challengers and state security forces. In this arena of interaction, called the protest-repression nexus, protests by resistance movements are met with repression by government forces that in turn generates more dissent by protestors.[39] This tit-for-tat strategy contributes to the emergence of violent contention.[40] As challengers leverage disruptive and conventional forms of protest, political authorities muster their own response. The response is typically a mixture of repressive measures and reform policies designed to demobilize the resistance movement by raising the costs of resistance but also detracting from the movement's appeal to its broader audience.

While the reforms occur in the hallowed halls of state legislatures, repressive measures play out on the streets as domestic security forces attempt to restore law and order.[41, 42] Broader factors in the environment also play a role, as some regimes are more willing than others to use harsh repressive measures.[43] Some policing will stay within the bounds of "negotiated management" that includes legitimate, legal, and conventional methods of containing challenges to political authority.[44] In other cases, however, policing devolves into a pattern of interactive violence that is a key factor leading to protest escalation.[45]

The protest-repression nexus emerges from the dynamics of a movement's requirement for tactical innovation and adaption. Because most resistance movements are less powerful than the states they confront, they rely on continual innovation to disrupt public order and to mobilize supporters. Meanwhile, security forces are also forced to innovate and adapt to counter the challenges to political authority. It is within the back-and-forth tactical interaction where transitions to violence emerge. In his analysis of the black movement in the US during the 1960s, Doug McAdam describes how black insurgent groups continually adapted tactically in response to interactions with southern white segregationists to sustain the movement's momentum.[46, 47] The movement's tactical innovations corresponded with peaks of movement activity. The tactical innovations included the bus boycott, the sit-in, the freedom rides, community campaigns, and finally the urban riot. After a tactical innovation, the movement's activity declined as opponents developed effective countermeasures. Each innovation generated a corresponding response from white segregationists, which included extralegal physical violence and intimidation, as well as legal obstruction carried out by law enforcement and government officials. As the efficacy of a tactic declined, the black movement adapted to other forms of protest that circumvented the countermeasures.[48]

Different Pace of Demobilization

Interactive violence also initiates internal dynamics that contribute to the adoption of violence. Repression can raise the cost of participation for many peripheral supporters, so they are often among the first to demobilize. As the moderates leave a movement, the remaining activists are the most "hard core" and the most likely to use violence, a mechanism referred to as the "different pace of demobilization." Often, government authorities also seek to negotiate or co-opt the moderates while repressing more radical components of a movement. As a result, upticks in resistance violence are paradoxically often first seen in a movement's decline as peripheral members or moderates leave the organization and the hard-core activists remain to take the reins of control.[49]

Different pace of demobilization occurs when state repression and negotiation raise the cost of participation for members or resolve pressing issues, leaving a highly committed, radicalized core that raises the likelihood of a transition to violence.

Internal Competition for Power

Interactive violence between a movement and security forces also impacts interaction among actors within a movement. Actors within a movement may be competing among themselves while also competing with an external adversary. In this regard, the strategies and tactics of a movement can be driven by factors internal to the organization, not just by external pressures. While resistance organizations are often treated as monolithic, homogenous entities, movements include a diverse field of actors. Movements may start out on the basis of shared interests and goals, but they rarely stay that way as differences emerge over ideology, strategies, tactics, and goals.[50] Moreover, most movements include organizations and individuals who are less willing to support a transition to violence (moderates) and those more willing to support such a transition (radicals). Competition for power among these subgroups can be a decisive factor in a violent, radical splinter group's departure from a larger nonviolent social movement. The splinter group then actively generates new opportunities for violence in the ongoing protest cycle.[51] Repression can instigate competitions for power between various actors within a movement as they struggle over whose tactics and strategies should be adopted. As Tarrow describes:

> Competition may arise from ideological conflicts, from competition for space, or from personal conflicts for power between leaders. Whatever its source, a common outcome of competition is radicalization.[52]

Nevertheless, the adoption of violence is generally an incremental process. At the outset, violence is largely unplanned and arises organically from predominantly nonviolent tactics, whether sit-ins, occupations, or protests. Spontaneous violence might emerge, for example, in clashes between police and demonstrators at a large protest. The small-scale violence, however, contributes to the development of small units within the organization that "specialize in increasingly extreme tactics" and "[build] up an armory for such tactics."[53]

Transitions to Violence Among Left-Wing Resistance Movements

In her research on left-wing opposition groups in Italy during the 1960s and 1970s, Donatella della Porta explains how radical splinter groups broke off from the main left-wing social movement that had until then adopted nonviolent strategies. Facing stiff repression from the state, segments of the left-wing movement disagreed about the most effective way to pursue their goals, whether through violence or nonviolence. Ironically, the disagreements led to physical violence between members that required the formation of self-defense squads for protection against other left-wing activists. The self-defense squads engaged in daily violent altercations with their political opponents in the state, other left-wing activists, and countermovement opponents on the fascist far-right. Individuals involved in the self-defense squad reported becoming slowly habituated to violence as the weapons of choice evolved from fists, to stones, then eventually to guns. Violent struggle against the Italian state peaked in the 1970s with the formation of the '77 Movement and the Red Brigades, radical leftist clandestine groups that practiced armed struggle.[54]

Political Outbidding

Another causal mechanism activated during protest cycles that contribute to the transition to violence is political outbidding. Political outbidding is a form of competition that arises between groups competing for scarce resources in the organizationally dense field of actors that often makes up a resistance movement. Similar to the competition for power mechanism previously discussed, disagreements between the groups arise over the appropriate strategies and goals to adopt. However, groups with similar goals at the outset also find themselves competing for sources of external funding, allies among the political elite, recruits, legitimacy, and media coverage, among other important resources. Political outbidding describes how competition among these groups contributes to violence by encouraging them to seek ways to differentiate their group from the others in the field. Outbidding is defined as "action-counteraction dynamics in which each side raises the stakes in response to the other."[55] One way to differentiate, of course, is through the adoption of increasingly violent tactics. As one group adopts violence, other groups are oftentimes forced to follow in an effort to "outbid" one another.[56]

Political outbidding is a form of competition that arises between groups competing for scarce resources in the organizationally dense field of actors that often makes up a resistance movement.

The adoption of violence occurs when radical flanks of a group use extreme claims to protect their community and interests. The radicals also accuse the groups that remain moderate of selling out, which in turn can spur moderate groups to adopt more radical, violent tactics in an effort to remain relevant. Political outbidding among groups seeking to differentiate themselves from the crowd helps to radicalize across the movement as it legitimizes and normalizes radical action. Whereas before violence was seen as extreme, it becomes increasingly viewed as a morally appropriate and righteous means of protecting a community's interests, values, and beliefs.

Countermovements to Irish Republican Insurgency in Northern Ireland

Gianluca de Fazio explains how political outbidding occurred in Northern Ireland among the many organizational actors in the 1960s civil rights movement that eventually gave way to a decades-long insurgency waged by the PIRA. External dynamics, such as state repression and the formation of a hostile countermovement, aided the efforts of some groups in the civil rights movement to radicalize. The civil rights movement in Northern Ireland first developed under a large umbrella organization, the Northern Ireland Civil Rights Association (NICRA), which challenged the discriminatory policies of the Protestant-dominated Northern Irish state that marginalized various sectors of society, especially the Catholic minority. The NICRA, while largely moderate, contained radical elements, particularly the student-led People's Democracy. While initially the groups cooperated, differences arose as the People's Democracy advocated transgressive tactics that sought to provoke and antagonize the Protestant-dominated police force and Protestant Loyalists who defended Northern Ireland's continued ties with Great Britain. In 1969, the People's Democracy arranged a march through a virulently Loyalist neighborhood in Derry, the second-largest city in the country, in an effort of "calculated martyrdom." The demonstrators were predictably mobbed and attacked by Loyalists who were aided by off-duty police. After the brutal attacks, which were widely televised, the People's Democracy gained a great deal of sympathy, and resources, from the Catholic population.

At the same time, moderates gained ground in conventional politics as civil rights leaders were elected to seats in the Northern Ireland Parliament in 1969. Mobilization dwindled as moderates in the civil rights movement stepped away from transgressive contention of mass protests and demonstrations to push their aims through legitimate political channels. In the environment of increasingly scarce resources, the People's Democracy began to emphasize its militancy to distinguish itself from the NICRA and other moderate groups in the civil rights movement. Its provocative tactics, such as the march in Derry, gained supporters, allowing the radical group to exercise more leadership over the movement. Under more radical guidance, and in an increasingly ethnically divided society, protests on the streets of Belfast and Derry became more violent as police, Loyalists, and Catholics clashed repeatedly. Civil disturbances gave way to ethnic riots and sectarian attacks. The People's Democracy continued to lambast the more moderate NICRA for its failure to produce tangible results or protect the Catholic minority. By 1969, the British Army was deployed to contain the explosive situation.[57] Splinter factions within the Irish Republican Army (IRA), citing the need for defense of the Catholic community, formed the PIRA in 1969. The PIRA, whose membership included radicalized members of the civil rights movement, openly advocated for armed struggle against Britain and the Northern Irish state in pursuit of reunification with the Republic of Ireland.[58, 59]

The radicalization after political outbidding also illustrates how a movement's claims shift alongside its tactics. The transition to violence in Northern Ireland was also accompanied by a shift in the claims made by the movement. At the outset of the civil rights movement, protesters advocated for limited political, social, and economic reforms within the Northern Ireland state. The radicalization in tactics coincided with a transition from these limited objectives of protesters to maximalist demands after repressive action by counterprotesters. Increasingly, radical components of the civil rights movement, including the People's Democracy, demanded the dissolution of the union with Great Britain and reunification with the Republic of Ireland. The increasing use of violence by both sides led to the transformation of demands, as state violence increased protesters' perceptions of the illegitimacy of the Northern Ireland state and led to boundary activation or an increase in us–them thinking, which in this case corresponded to an increasing salience in ethnic differences between Protestants and Catholics.[60] In this regard, relational dynamics also play a role in leading to an escalation of claims made by resistance actors.

STRATEGIES AND TACTICS

The previous section looked at how relational dynamics contribute to the adoption of some strategies and tactics over others, particularly in the transition from nonviolence to violence.[61] Other approaches in political science look at the question from different perspectives, but generally in reference to groups already engaged in an insurgency or civil war.[62, 63] Environmental conditions and resistance movement leaders' early choices impact future behaviors or limit the range of strategic and tactical choices available to them. Using theory developed in organizational studies, political scientists analyze how a movement's organizational structure influences its decision-making and behavior, whether in terms of operational capacity or ability to strategically plan for the movement as a whole. Decisions on organizational structure are often based on the delicate balance between command and control on the one hand and secrecy and operational effectiveness on the other.

Leaders need to make decisions about the best way to organize but also about how to resource their operations. Violent insurgent groups use a variety of methods to finance their operations, from taxes on local communities to drug trafficking and natural resource exploitation. Just like a group's organizational structure, the methods a group uses impact how the group relates to the host population it purports to represent. Predatory financing methods are associated with higher levels of indiscriminate violence and civilian deaths, leaching popular support for the group.

Many, but not all, violent insurgencies form because of struggles related to territory. Some organizations are successful in gaining control of territory, while others are not. Territorial control, or lack of it, has a significant impact on an organization's operational tactics. Some political scientists even conjecture that terrorism is a tactic used almost exclusively by clandestine organizations that lack such control. Moreover, groups that do have territorial control are able to establish shadow governance activities that provide opportunities for population control and increased legitimacy among the host population.

Organizational Structure

Insurgent groups generally adopt one of three main types of organizations: hierarchies, networks, or edges.[64] A classic military unit has a strict hierarchy where authority is established according to one's level and position. Hierarchies are contrasted with network structures, which are organically evolved sets of relationships between individuals. Authority in a pure network comes from personal relationships, reputation, and to some extent network position. In this type of structure, more well-connected individuals may exert more influence. David Alberts and Richard Hayes popularized a third option called an "edge" structure.[65] An edge has clearer vertical lines of authority than a network. Authority is partially established by position, but it is designed to be much flatter than hierarchies. This allows more cross-cutting connections for information flow, distributes more decision-making authority to individuals, and is more flexible in allowing ad hoc teams to form and focus around specific problems.

The choice between hierarchy, network, and edge organizations involves trade-offs, with each type of system having different strengths.[66, 67] Organizational research emphasized the importance of matching organizational forms to specific tasks and situations:

- *How important is speed of response?* Hierarchical command structures involve delay as requests move up and down a chain of command and bottleneck when managers have to prioritize some decisions at the expense of others. Flatter systems typically gain speed but sacrifice reliability and quality control.

- *How important is specialization?* As an organization matures, it begins to set up specialized functions such as media production or weapons training. If it does not, it risks wasting resources on redundant efforts.

- *How costly are mistakes?* Private corporations can tolerate a great deal of variance in quality for the sake of novelty and freshness; likewise, a newly formed insurgency might highly value surprise at the cost of the occasional operational disaster. An established automobile manufacturer may have much less tolerance for manufacturing defects; likewise, a late-stage insurgency trying to establish legitimacy may have much less tolerance for accidental civilian deaths or public relations disasters.

- *How important is unity of command?* Effective hierarchical command structures maximize consistency. In a mature insurgency, coordination of units and military, political, and communications actions can be crucial.

- *Are there cultural reasons to favor centralization?* A hierarchy may be preferable if an organization's membership strongly favors it. This could be the case in a resistance movement that recruits heavily from ex-military personnel.

Another key decision resistance leaders make is between regional and functional structures. Regionally based groups locate all functions in a specified area. This organizational type is especially responsive to local conditions. By contrast, functionally based groups have specialized support groups within a centralized command. Leadership dispatches specialists to other groups as needs arise. Structure also strongly affects an organization's ability to learn and adapt. Groups with regionally based structures are generally better at perceiving and adapting to local conditions. Those in functional structures, where specialists in an area (e.g., military strategy) mostly interact with other specialists, may attain a higher level of professionalism in that specialty and be able to take on more complex tasks.

Decisions regarding organizational structure appear to make a difference in the functioning of insurgent groups.[68] Violent groups that adopt a vertically integrated functional structure, and thus have specialized units dedicated to violence, commit more lethal attacks and sustain violence over longer periods of time. Although networked organizations enjoy operational advantages because of their greater flexibility and autonomy, hierarchical organizations are able to produce more large-scale sustained violence as a result of three benefits of hierarchy: (1) clear lines of command and control that coordinate efforts by different parts of an organization, (2) accountability between functional units and central authority, and (3) specialization within units.[69] The Basque separatist group Euskadi Ta Askatasuna (Basque Homeland and Freedom, or ETA) in Spain reorganized itself between vertical and flat structures several times and demonstrated more violence during periods of vertical organization.[70]

Each of these benefits in turn facilitates processes that enable hierarchies to sustain higher levels of production of violence. Clear lines of authority in hierarchies enable greater leadership agenda setting. The relationship between the agenda setter and subordinate units is clear and unidirectional.[71] Networked organizations have multiple actors fulfilling the same functions, making them more flexible and resilient. At the same time, it can lead to unclear accountability and competition over agenda control.[72] Additionally, hierarchical groups such as Hamas and Lebanese Hizbollah have specialized units that focus on producing specific outputs, such as violence, public goods (i.e., health care and education), and political activity (i.e., standing candidates for political office). As a result, those carrying out violence within hierarchically organized insurgent groups are specialists able to sustain higher incidences of violence that are also more lethal than nonspecialists.[73]

Organizations spend considerable resources managing internal knowledge. These initiatives may involve creating information systems to ease sharing of information across divisions. While in business organizations, open, cross-cutting information sharing is preferable, insurgent groups operate in conditions of secrecy. As a result, communication must be carefully managed. The need for secrecy and security can trump all other concerns, particularly in the early incipient phase of resistance when the group is especially vulnerable to interdiction.

The demand for secrecy means that compartmentalization is often a critical organizational priority. In compartmentalized organizations, members are divided into cells where they have little knowledge or contact outside of the cell. If one member is compromised, the arrest does not have cascading effects across all the cells, endangering the larger organization. Usually only the leader has contact with the outside and relays instructions or requests through a single contact outside of the group. In some cases, even the leader may not have regular personal contact with cell members.

While it satisfies the demand for secrecy, a networked, decentralized, or compartmentalized organizational structure can produce suboptimal outcomes. Jacob Shapiro calls this the terrorist dilemma.[74] The need to stay covert forces insurgent groups to adopt a networked structure, where local cells are empowered to make decisions on all operational and administrative matters, including spending, tactics, and targets. However, the decisions cells make may differ from the preferences of leaders and could ultimately harm the interests of the group.

The potential tug-of-war between resistance leadership and autonomous cells produces principal–agent issues and agency losses. A principal, in this case a leader within a terrorist movement, must delegate authority for carrying out tasks to an agent, the terrorist operative. Agency losses arise when the preferences of agents on matters related to spending, tactics, and target selection differ from those of the delegating principals in charge. An exchange between Abu Hafs al-Masri, an al-Qaeda military commander in the late 1990s, and Abu Khabab, a member of Egyptian Islamic Jihad who began working closely with al-Qaeda in the late 1990s, illustrates the issue of preference divergence and agency loss. In the exchange, al-Masri, the principal, objects to the spending activity of Khabab, the agent, claiming the latter failed to properly submit claims for his travel expenses with the al-Qaeda accountant, took too much sick leave, and overspent on an air-conditioning unit.[75]

Issues of preference divergence also arise on more important matters of strategy and tactics. In particular, agents may carry out violence that principals view as counterproductive. For instance, in 2005, Ayman al-Zawahiri, then the second-in-command of al-Qaeda, wrote a letter to Abu Musab al-Zarqawi, the leader of al-Qaeda in Iraq, chastising him for carrying out beheadings and urging him to consider other less graphic methods:

> Among the things which the feelings of the Muslim populace who love and support you will never find palatable—also—are the scenes of slaughtering the hostages . . . we are in a battle, and that more than half of this battle is taking place in the battlefield of the media. . . . And we can kill the captives by bullet. That would achieve that which is sought after without exposing ourselves to the questions and answering to doubts. We don't need this.[76]

Counterproductive violence is also attributed to the covert structure terrorist groups must adopt to survive. The necessity of avoiding government forces means that terrorists lead isolated lives, which may in turn make them ill informed regarding the potential impact of their actions. Shapiro noted that this dynamic was present during the Islamist challenge to the rule of President Hafez al-Assad in Syria from 1979 to 1982, when local cells repeatedly conducted attacks that the outside leadership opposed, given their knowledge of the broader political situation.[77]

Leaders of terrorist groups face difficult choices while trying to minimize agency losses that arise from these principal–agent issues. The organizational practices that could minimize agency losses also expose terrorist organizations to a greater scrutiny by security forces combatting them. For instance, greater bureaucracy and paperwork are ways of acquiring information on the activity of agents.[78] Such oversight, however, generates additional communications activity, which in turn can pose a security risk. Al-Qaeda in Iraq kept very detailed and centralized records of its financial activity, which proved useful for counterterrorism efforts once the records were captured by American forces in 2007 and 2008.[79] The terrorist dilemma is often inescapable. It illustrates how choices regarding organizational structure lead to inevitable tensions and contradictions within violent resistance movements that impact their activities and use of violence.

Organizational Structure and Resistance Movement Resilience

Whether violent or nonviolent, resistance activities may sometimes be carried out by a host of organizations rather than by a single entity. Some researchers, therefore, speculate whether the network structure of the overall resistance movement helps explain the sustainability of resistance activities over a period of time. Resilience is especially important in nondemocratic regimes under which political opposition is generally not legal or faces higher and more severe rates of repression.

In a study of the resistance against the communist government in Poland in the late 1960s and late 1970s, Maryjane Osa argued that the resistance movement in the latter period was more resilient in the face of government repression.[80, 81] During the previous period, protest occurred in "islands of opposition" carried by one class, whether it was the students and intelligentsia in 1968 or the workers in 1970 and 1976. By 1979, the stage was set for the "tacit alliance of workers, intelligentsia, and Church," which was to grow into the Solidarity movement.[82] The communist government was able to break protests by activating existing social cleavages between these groups.[83, 84] The Solidarity movement that emerged in the later period was more resilient because the constellation of organizations that made up the resistance movement in the late 1970s was organized along multiple foci. The coalition building evident among opposition networks during the late 1970s meant that a broad swath of society was represented in the Solidarity movement. Extensive coalition building was enabled by the flexible civic discourse predominant in the movement that allowed a diverse set of individuals to join without confronting ideological barriers, such as those associated with strict religious content.

While the Solidarity trade union was a focal point of the resistance movement after the group's emergence in 1980, it was embedded in a broader interlinked network that featured other important organizations, including Klub Inteligencij Katolików, a club for Catholic intellectuals; Komitet Obrony Robotników, a workers' defense committee; and Towarzystwo Kursów Naukowych (Society of Scientific Courses), a group organized by academic and cultural figures that hosted salons in private apartments on a rotating basis to promote the discussion of politically sensitive topics.[85] This dispersed network structure featured one prominent focal point, Solidarity, but also other subordinate yet influential foci. This facilitated the flow of information and resources, enabling support for more sustained protest activities in the early 1980s.[86, 87]

Framing

Chapter 3 discusses the difficulties that resistance movements confront in mobilizing populations to actively participate in the movement or passively support the movement. The back-and-forth, interactive dynamics between the resistance movement and its target audience, whether other like-minded organizations, the media, political elites, sympathizers, bystander publics, and adversaries, are a form of social interaction. These interactions, and indeed most forms of human interaction, are conditioned by the identities of the participants in the interaction. For instance, the interactions that occur in a classroom setting are based on the identities of teacher and student which each prescribe expected responsibilities and behaviors.

There are numerous classes of identity, including personal, social and collective. At any given time, an individual has multiple identities—some that overlap, complement, or even conflict with one another.[88, 89]

133

The salience of a particular identity at a given time is dependent on the interaction. Personal identity forms the most basic identity and can be the building block for other identities. Personal identity is based on an individual's life experiences, especially the experiences that people single out as important or significant in their lives that shape personal attributes. A social identity contains both role and categorical identities. As indicated in the example of the classroom setting, one component of social identity is role identities. Role identities are based on the roles that people play or assume in the course of their lives. The roles might be professional, as in a soldier or teacher, but also familial, such as a mother, sister, or daughter. In addition to a role identity, social identity also contains categorical identities. These are familiar forms of identity associated with specific social categories, including gender, sexual orientation, ethnicity, or nationality. Notice that categorical identities have been notably influential in politics, called identity politics, in the past several decades.

The last class of identity, collective identity, has been singled out as especially important to resistance movement mobilization. It is theorized to be a key stepping stone between motivating people to transition from observers to supporters or participants. A collective identity is an identity of a group that generates the sense of "one-ness" or "we-ness" based on real or perceived shared attributes. Oftentimes, collective identity is based on a shared social identity such as ethnicity or religion, but the mere presence of these social identities does not immediately translate to a collective identity. Instead, the development of a collective identity often emerges as groups interact with other conflicting groups. One example is the rally-around-the-flag effect of terrorist incidents such as 9/11 that enhance a sense of collective identity surrounding nationality. Collective identity is critical for resistance movements because coming together requires that people share the same idea about what separates them from other groups, the grievances the group needs to address, the most appropriate solutions to those grievances, and a vision for the future. Together, these shared interests form a collective identity which in turn is based on the alignment of people's social and personal identities with it. The formation of a robust collective identity is a crucial first step toward collective action.

As a result, one of the most important strategies that resistance leaders can focus on is called identity work. Identity work is described as the "range of activities in which individuals and groups engage to give meaning to themselves and others by selectively presenting or attributing and sustaining identities congruent with their interests."[90] Identity work for resistance movements occurs primarily through framing.[91, 92] Erving Goffman first defined frames as "schemata of interpretation" that enable people "to locate, perceive, identify and label" events that they experience or are brought to their attention.[93] More simply, a frame represents a worldview through which events and concepts are interpreted. Resistance movements strategically leverage and actively construct these worldviews and interpretations to encourage their target audience to adopt the group's collective identity.

A broad class of frames, called collective action frames, are singled out because they are crucial in forming collective identities and thereby motivating collective action on behalf of a resistance movement. Collective action frames are "*action-oriented* sets of beliefs and meanings that inspire and legitimate the activities and campaigns of a social movement organization."[94] The frames perform an interpretive function by condensing the "world out there,"[95] especially in ways that mobilize potential supporters and demobilize antagonists.[96]

Collective action frames perform three core framing tasks: diagnostic framing, prognostic framing, and motivational framing.[97] The first involves identifying the problem and its victim and attributing the problem to responsible actors and causes. The second entails specifying a proposed solution and a strategy for carrying out corrective action. The third provides a rationale for engaging in remedial collective action, including an appropriate vocabulary of motive.[98]

Each of these concepts can be fruitfully applied to various resistance movements. Many diagnostic frames are injustice frames that identify a population as a victim of injustice at the hands of another. The diagnostic frame of the PLO encompassed a narrative that focused on the loss of Palestinian land to Jewish settlers, the displacement of much of the original Palestinian population, and the establishment of a Jewish state. In the case of the Shining Path of Peru, the diagnostic frame centered on the arrival of the Spanish and the downfall of the Inca Empire in the sixteenth century as the cause of the current economic and social misfortune of the indigenous population of Peru.

In some cases, history provides an "original sin" that will inform the initial narrative of an insurgency. It may also provide the set of actors as well as their corresponding actions and group grievances that populate the discourse of a diagnostic frame. The grievances that informed the diagnostic frame of the Shining Path originated from an event more than four hundred years in the past. In the case of Shia revolutionaries in Lebanon and Iran, the original sin occurred 1,300 years before with the death of Husayn ibn Ali in 680 CE. In other cases, history is less influential. The grievances that motivated the Movement for the Emancipation of the Niger Delta in Nigeria were based on the actions of the Nigerian government and not on, for example, the actions of the British during the colonial era.[99]

Another important aspect of diagnostic frames is that they are highly contested. Various groups and individuals belonging to resistance movements debate among themselves about the nature of the problem (i.e., the diagnostic frame) and, consequentially, what is to be done about the problem (i.e., the prognostic frame). For instance, a variety of Palestinian groups emerged to challenge Israel. Some, such as the Fatah-led PLO, eventually came to recognize Israel and, consequently, developed a diagnostic frame that centered on the Israeli occupation of the West Bank and Gaza Strip. Other groups, such as Hamas, took a maximalist stance by refusing to recognize Israel, so their diagnostic frame is notable in that it identifies the problem as Israel's existence rather than simply its occupation of a portion of land between the Jordan River and the Mediterranean Sea. These differences between diagnostic frames lead to differences in prognostic frames (i.e., negotiations as opposed to perpetual resistance against Israel), and indeed such differences have led to internecine fighting between Hamas and Fatah.[100]

A group's motivational frame represents a call to arms.[101] Motivational frames emphasize the need to address historical and modern-day grievances. However, sometimes the most effective motivational frames are those that equate actions with affirmation of identity, personal significance, and even personal salvation. During the 1979 Iranian Revolution, Iranian clerics and jurists adeptly manipulated motivational frames. In speeches and Friday congregational sermons, leaders extolled the notion of martyrdom by equating self-sacrifice with an honorable death in support of the revolution. The exemplary martyrdom of the Shia imams at the advent of Islam provided the blueprint. Martyrdom in defense of the Islamic government was therefore imbued with deep personal significance because it offered a direct path to personal and eternal salvation in the afterlife and, in fact, a privileged position in the ever after right next to the pantheon of martyred Shia imams.[102]

The frames a resistance movement adopts need to resonate with the target audience to be effective. Resonance occurs when the frames graft, expand, or reconceptualize existing narratives that are meaningful or important to the target audience. The framing concept that describes this critical task is frame articulation, defined as "the connection and alignment of events and experiences so that they hang together in a relatively unified and compelling fashion."[103] In this way, messaging by insurgents typically incorporates existing themes and ideas. Despite the audience's familiarity with the material, it becomes novel when woven together in new ways, creating a compelling story in which causality, attribution, and corrective action are clear and resonate with the target audience. The result is the construction of a reality and meaning that may resonate with the segment of the population that an insurgent group wants to mobilize. The novelty of the Shining Path's message was its ability to tie together, through diagnostic, prognostic, and motivational frames, historical and current grievances in a way that still resonated with its audience.[104]

Lastly, frame transformation is apparent in most resistance narratives. Frame transformation is the "changing [of] old understandings and meanings and/or generating new ones." When history provides much of the raw material for a comprehensive narrative, crafters use frame transformation to reinterpret the past to encourage mobilization. During the 1979 Iran Revolution, clerics strove to overturn a quietist tradition in Shia Islam that shunned political activity for a millennium.[105] The clerics used Friday sermons to overturn this passive conception of Shia Islam and to argue instead for its revolutionary nature.[106]

Territorial Control

In the 1960s, author Robert Ardrey published a work, *The Territorial Imperative*, discussing human territoriality, which is the instinct to acquire and defend land.[107] He explained territoriality as a fundamental human instinct, shaped through the millennia by evolutionary processes, motivating the behavior of even modern humans. Indeed, territory, either the desire to acquire it or defend it, remains a significant factor driving interstate and intrastate violent conflict. Territorial disputes were a factor in 65 percent of conflicts involving two parties between 1816 and 1945, and for conflicts after 1945, this figure rose to 72 percent.[108] Territorial disputes also play a prominent role among civil wars; half of them since 1990 involved ethnic groups seeking statehood or greater autonomy.[109]

As its importance as a motivator of violent conflict suggests, territorial control is a game changer in a number of ways. The extent of territorial control by armed actors in a conflict is an important intermediary affecting how the group interacts with the population, the type and intensity of violence needed to counter its rival, and its ability to mimic a state through shadow governance activities. Territorial control enables an armed group to transition from an underground, clandestine organization with circumscribed operational capabilities to a mature insurgency with a robust armed component. Territorial control also offsets the inherent limitations of the terrorists' dilemma previously described as formerly clandestine groups are able to operate more like state governments than shadowy networks. J. Bowyer Bell, a social scientist and longtime fellow traveler among clandestine armed groups, cites the distinction between the licit and the illicit as a key factor limiting the rebel ecosystem:

> The two most important obstacles to an effective underground . . . are the necessity for cover and the legitimacy of the opponent that makes the movement not only covert but also illicit. . . . The more secret an organization, the more inefficient, and absolute secrecy assures total chaos.[110]

Most armed groups, therefore, struggle for liberated zones and "no-go" areas that can help them acquire the attributes of their legal, overtly organized state opponents.[111, 112]

Territorial control has a strong impact on the range of tactics available to violent resistance movements.[113, 114] At the least, groups with territorial control have camps and bases within the border of the country in which they operate; at the most, the insurgents replace the state as the de facto sovereign in the territory, establishing a shadow or parallel government. By contrast, groups that do not control territory are forced into hiding most of the time. Groups with territorial control can establish camps, train recruits, and use heavy equipment, activities that in turn enable specific types of tactics, such as raids, skirmishes, ambushes, small-scale battles, and seizure of villages.[115] Due to logistical and organizational limitations, clandestine groups cannot use these tactics and instead tend to use other methods, including assassinations, selective shootings, bank robberies, and bombings.[116]

Evidence supports the increased operational range of armed groups with territorial control. An analysis of the tactics used by 122 armed groups shows that groups with territorial control carried out more attacks on facilities, typically requiring large teams and the occupation of space. Clandestine groups, by contrast, more often carried out assassinations and bombings, tactics that do not require territorial control.[117, 118] In the case of Hizbollah, before the group established territorial control in South Lebanon and Beirut in the late 1980s, its tactics consisted mainly of bombings, assassinations, and kidnappings. Yet after gaining territorial control, the group increased its attacks on facilities, and this became its main tactic throughout the 1990s.[119] The evidence correlating clandestine organizations without territorial control with terrorist-type attacks has even led some researchers to argue that terrorism should be associated primarily with these groups. While insurgent groups that control territory can and do use terrorism, the tactic is generally deployed in areas in which the group does not exercise full control of the territory.[120]

Armed groups that exercise territorial control are also able to wield violence more precisely against rival collaborators. Kalvyas, in his control-collaboration model, explains how armed groups with territorial control can take advantage of selective violence in the place of indiscriminate violence, which allows the group to maintain greater legitimacy and popular support.[121, 122] The model is intended to explain patterns of violence in civil war that might otherwise seem chaotic or irrational. A crucial assumption in this model is that in armed struggle, one of the most important dynamics is that between an armed actor, such as an insurgent group, and the population.

Insurgents require some level of explicit or tacit support from this population to prevail against their stronger state rival. If successful armed groups need to maximize popular support, the corollary is that the group also needs to minimize defection. Here, defection refers to any support an individual offers to an armed group's rival. Armed groups can use numerous strategies to minimize defection, but when territorial control is contested between insurgents and government forces, coercion or violence is more likely to prevail. However, for violence to be optimally effective, it needs to be directed toward the individual actually engaging in defection. This sort of violence is referred to as selective because it targets the individual directly responsible for the sanctioned behavior.[123]

Armed groups, however, struggle to gather the intelligence necessary to use selective violence, causing them to rely instead on indiscriminate violence. Other civilians in the community are the most likely source of accurate intelligence; there is an asymmetry of information because civilians have the information that the armed group requires. When an armed group does not have accurate intelligence, it must rely on indiscriminate violence to minimize defection. This means that whole categories of people are targeted, rather than an individual. The armed group might suspect that males of military age are the most likely defectors and thus indiscriminately kill all individuals fitting those criteria, or the group might suspect members of a certain ethnic group, or village, and so on. Regardless, the result of indiscriminate violence is usually a terrorized population holding deep-seated grievances against the responsible party.[124]

The information asymmetry confronting armed groups attempting to minimize defection is best mitigated through territorial control. When armed groups have some territorial control, they have improved access to civilians' information on defectors. In the control-collaboration model, one of the primary deterrents facing civilians wishing to provide information to an armed group is fear of retaliation from supporters of a rival group. If you report the defection of your neighbor, you worry that your neighbor's family will in turn report you to the rival forces, which puts your life in considerable danger.[125]

Kalyvas observed that territorial control exists on a five-gradient spectrum. In the middle, a territory is contested by both parties in the conflict, with neither having a decisive advantage over the other. On either end, the armed group or the rival state has full control of the territory in question. The remaining two gradients convey fragmented control of the territory, when either the armed actor or the state has predominant, but not complete, control of the territory in question. Because the presence of the rival actor is minimized in areas of fragmented control, civilians are more likely to offer intelligence on defection as the chances of retaliation are very limited. The control-collaboration model, based on levels of territorial control, predicts when armed groups are likely to use indiscriminate or selective violence. Indiscriminate violence is most likely in highly contested zones, while selective violence is most often found in areas of fragmented control. The incidence of violence is nearly absent in areas fully controlled by either the armed actor or the state. Because control of the territory is not contested, further violence is largely unnecessary.[126]

Varying levels of territorial control by either state or non-state armed actors, then, is a pattern that explains why violence in some areas is so bloody and intense, while it is less devastating in others. In a conflict between an incumbent state and an insurgent group, it is usually the state that first unleashes campaigns of indiscriminate violence. The campaigns, whether called mopping up, scorched earth, or cordon and search, are signals that the regime recognizes its fragmented control over its territory but lacks sufficient intelligence to accurately target the insurgents amidst the civilian population.[127]

The Control-Collaboration Model in the Greek Resistance

The empirical evidence for the control-collaboration model comes from extensive quantitative and qualitative research of the interactions between the German occupation forces in Greece and the British-supported Greek resistance against the occupiers, the National Liberation Front (Ethniko Apeleftherotiko Metopo, or EAM) and the Greek People's Liberation Army (Ellinikós Laïkós Apeleftherotikós Stratós, or ELAS), during World War Two. In September 1942, German victories were at their height. The German war machine was in high gear, and the summer campaigns were still going strong. One of the major supply routes for General Rommel's forces in Africa was from Germany through Greece. Britain and the Allies hoped to cut off General Rommel's supply route through Greece with the assistance of the EAM and ELAS guerrillas, despite the communist ideology the groups espoused.

In the hills and lowlands of Greece, as rural villages shifted between the Germans and the guerrilla partisans, patterns of indiscriminate and selective violence emerged. Overall, the EAM and ELAS insurgents benefited from extensive local networks that provided good intelligence. They relied primarily on selective violence, accounting for about 68 percent of the violence they perpetrated in the Argolid, a region in southern Greece.[128] The Germans, an occupying force who did not have the benefit of local networks at first, relied heavily on indiscriminate violence, also around 68 percent, in the same region.

Overall, Kalyvas found a great deal of support for his control-collaboration model. Nearly all of the indiscriminate violence in the Argolid occurred in zones 2 and 4 of territorial control, perpetrated by the political actor that had the least control in the region. Conversely, selective violence was most commonly used in zones 2 and 4 by the political actors that enjoyed predominant control in those regions. As expected, little violence was experienced in villages that fell under the complete control of either the occupying German or Greek resistance forces. Within the Greek village Maledreni, which was in zone 4, insurgents relied on selective violence to maintain predominant control. When villagers denounced German supporters to the ruling council, village leaders were generally able to sift charges with merit from those made for reasons of personal revenge. When German forces began a series of successful incursions against the village, attempting to gain control, they relied on indiscriminate violence to root out resistance supporters, massacring scores. Likewise, in another Greek village, Heli, in zone 2 of territorial control, German occupiers were also able to rely on selective violence to further consolidate control. A surprising number of the local Albanian-speaking population was ready to denounce resistance supporters as the EAM and ELAS had treated the villagers poorly when they held control for nearly two years.[129]

Territorial Control and Public Services

With the exception of more predatory groups, insurgent groups often consciously adopt a shadow governance strategy to shore up popular support in their territories.[130, 131] Popular support may be active, as in the case of undergrounds or auxiliaries who perform activities short of combat, such as subversion, running safe houses, storing weapons or supplies, or providing intelligence.[132] Shadow governments frequently make it easier for these components to perform these activities. The remaining mass base forms the group of more passive supporters.

Shadow governments are formal or informal non-state organizations that strategically leverage governance activities to fulfill operational objectives in relation to population support and control.

Insurgent groups vary not only in whether they pursue popular support as a strategy but also in how they go about capturing popular support. Insurgent groups must capture territory before they can initiate governance activities that cultivate popular support, and many groups engaged in armed struggle find this to be a difficult and unfulfilled endeavor. Some groups that ideologically advocate popular support as a strategy may also find that the practical realities of conflict preclude them from following through on such a strategy.[133] Groups that do seek popular support adopt different tactics to achieve their objectives. Some groups opt for voluntary support, while others use more coercive measures. Especially for those groups seeking voluntary support, shadow governance activities often play a large role.

Insurgent groups that seek popular support and establish shadow governments face numerous challenges. Shadow governance requires scarce resources and incurs risk. Those encouraging voluntary support need to weigh the importance of governance activities against military goals. When faced with increased military pressure or infringements in controlled territory where governance occurs, insurgents can be forced to choose between protecting civilians and military survival.[134] When its survival was uncertain because of increased military pressure, the National Resistance Army (NRA) in Uganda, despite its previous commitment to popular support and civilian participation in its governing structures, was forced to abandon its territory and halted all shadow governance activities until its position improved.

Legitimacy is described as "generalized and normative support" for an incumbent authority.[135] The term *generalized* here means support that extends beyond approval of specific policies on particular issues, while the term *normative* refers to support that is not based on calculations of individual interest. There are numerous foundations for the legitimacy of political authority: tradition, as in the case of monarchies; the charisma of individual leaders; and legal-rational procedures, such as democratic elections, governing the selection of leaders.[136] The latter is process based, as legitimacy is based on the widespread acceptance of procedures for elevating individuals to positions of authority.

Legitimacy is the generalized and normative support for an incumbent state authority among the relevant population.

Maley notes that another basis of legitimacy is known as "social-eudaemonic," which means that the legitimacy of a government is based on its performance in meeting the demands of a population. In this case, a government is viewed as legitimate if it is seen as successful in providing a variety of public goods, including health care, education, sanitation, security, justice, and economic development. Another form of legitimacy is goal-rational legitimacy, where leaders' authority to rule is based on the desirability of the goals they pursue. The absence of this form of legitimacy can be seen in various Sunni regimes in the Arab Middle East, as groups such as al-Qaeda and ISIS have waged violent insurgencies and used terrorism to overthrow governments regarded as insufficiently Islamic. Legitimacy may be based on communitarian nationalism, as was the case with the LTTE, who waged a decades-long insurgency against the government of Sri Lanka in an effort to carve out a Tamil homeland in the island nation.[137] Legitimacy may also be based on the identity of the rulers, as was the case in postwar Iraq, as Sunnis withheld legitimacy to various post-Hussein governments in part due to bitterness that centuries of Sunni rule had come to an end following the US invasion in 2003.

The provision of public services can therefore be viewed as an effort to build legitimacy along social-eudaemonic lines. The tactical use of governance activities to influence perceptions of legitimacy or civilian behavior or to fulfill operational objectives is captured by the term *shadow governments*. As the term implies, shadow governments are formal or informal governance activities sometimes operating in tandem with those of the incumbent state government. Because of the predominance of the state as the accepted legitimate form of political organization, shadow governments mimic the attributes and functions of the nation state and, in effect, represent a "counter state."

Insurgent groups implement shadow governments in pursuit of a number of objectives. Oftentimes, and as implied by the preceding discussion, shadow governments are a reflection of the important objective to legitimate the authority of the insurgent group and gain popular support. However, insurgent groups may also use shadow governance activities to undermine the official government or to extract crucial resources. Shadow governments differ from one another in a number of ways, including their institutional complexity, effectiveness, and objectives. A number of factors are thought to account for these variations, including the political and military context, as well as a group's internal dynamics and objectives.

The provision of public services depends on the degree to which an insurgent group controls territory. The NRA, operating in Uganda, offered a series of services, including health care and security, to the civilian population in its liberated areas. As its hold on those areas deteriorated during its campaign against the incumbent government, the NRA evacuated civilians to safe pockets in the Luwero Triangle while still encouraging civilians to maintain the democratic village councils it had established in its safe areas. Eventually, as its position became more precarious, the NRA was forced to terminate all ties with the civilian population to free the group from allocating resources to civilian defense. The NRA demanded that civilians leave the war zone, and it only resumed governance activities when its military position vis-à-vis the incumbent government improved considerably.[138]

Shadow governments may provide a number of social services, such as building roads and telecommunications networks and providing education and health care. Lebanese Hizbollah has been especially effective in this regard. The group's Social Service Section used half of Hizbollah's 2007 budget for social services, which were delivered to the group's mostly Shia constituents.[139] The section is divided into six sub-groups supporting various needs of the community, from reconstruction, to providing for the families of martyrs, to women's welfare, to education. Hizbollah's social service efforts, such as the reconstruction of homes and structures damaged by the 2006 war with Israel, far outstrip those of the Lebanese state, which has done little to improve infrastructure in Shia neighborhoods since the 1900s.[140] Hizbollah's efforts, largely financed by Iran, have reaped handsome rewards in political and military clout in the Levant.[141]

Insurgents need to generate revenue to pay for the public services which they often accomplish through taxation. The creation of a legitimate state is "intimately bound with the creation of fiscal institutions that are acceptable to the majority."[142] In come cases insurgents may be more effective at gathering taxes than the government itself. Local populations may prefer the tax efforts of insurgents. Sympathetic populations may prefer to pay rather than evade taxes, as was the case during the Kosovo insurrection when the Kosovo diaspora in Germany contributed funds through a well-organized, if informal, payroll tax. Even populations expressing little support for insurgents may prefer insurgent to government taxes, particularly if the group is perceived as less predatory and offers more security than the government.[143]

Shadow governments and insurgent groups gather funds in a variety of ways, whether through taxation, voluntary contributions, extortion, kidnapping, or other criminal activities such as drug trafficking. During the Cold War, insurgencies frequently received funds from outside states. Today, most groups need a more diversified revenue stream, leading to an increase in the importance of control of "revenue-generating regions" and the potential for greater victimization of civilian populations.[144] The National Union for the Total Independence of Angola (União Nacional para a Independência Total de Angola, or UNITA) faced challenges after its primary source of income, foreign assistance, evaporated after the end of the Cold War. Afterward, the group's revenue-generating strategies transformed, relying heavily on territorial control of diamond-rich areas, as well as taking advantage of other commercial activities in the group's territory. Combined, these strategies generated as much as five million dollars a month. Despite an agreement to move toward centralized government, when government forces encroached on the diamond-rich territories, UNITA returned to violence.[145] Extraction of resources from a target population is, for some insurgent groups, the predominant objective of its governance activities.

In some cases, an incumbent government may work with officials from an insurgent organization to provide public goods to a host population. The Sri Lankan government provided services to the local population in parallel with the LTTE. The parties worked with each other, through a hybrid administrative system, to provide various social welfare services to the population under LTTE control. This was the case given the history of strong state institutions that had penetrated deeply into Tamil society by providing a substantial amount of public goods, including food subsidies, health care, education, and subsidized transportation.[146] The LTTE established an extensive apparatus for providing various social welfare services, including a police force and a court system, as well as ministries for health, education, finance, justice, and economic development, and both rebel and government officials developed a working relationship to coordinate the provision of aid to areas under the control of the LTTE.[147] In LTTE territory, a district-level official called a *porupalar* appointed by the group worked with a government official to ensure that the government provision of public goods was in accordance with LTTE policy.[148] Typically, government officials in rebel territories were ethnic Tamils who were acceptable to the LTTE and were sympathetic to the needs of the local population, and they usually had positive working relationships with their LTTE counterparts.[149]

The LTTE and the Sri Lankan government benefited from this arrangement. The government continued to provide welfare services to residents in LTTE territory because it still wanted to maintain even a tenuous link to a population it claimed to represent. Its abdication of the provision of public goods may have led the LTTE to develop an even more extensive and capable civil administration, thereby bolstering its image as a Tamil government-in-waiting for the region.[150] The LTTE also benefited from this arrangement, as it did not have to allocate a greater share of scarce resources to provide public goods, thus freeing up resources for combat activities. This arrangement also contributed to the tax base of the region under its control, as the government continued to pay the wages of government teachers, doctors, and other professionals in the health and education sectors in the region.[151] The LTTE imposed a tax of 12 percent on government workers, so in effect central government funds used to pay civil service personnel working in areas controlled by the insurgent group helped finance the LTTE. The government was aware that state funds were helping to finance the LTTE through taxes on civil servants but authorities regarded this cost as a necessary price to maintain a link to people living in insurgent territory.[152]

Resourcing

Resistance movements cannot form, mature, or act without resources.[153] It is conceivable that a spontaneous resistance movement might emerge from public dissatisfaction, a protest march, or a riot, but sustaining and growing a movement into something that will effect change requires time, patience, and, above all, resources. However, more than a means to an end, the methods insurgent groups use to obtain resources have a measurable impact on the strategy and tactics of a movement.

Depending on their activities, resistance movements may need money to meet a variety of expenses, including the purchase of arms, ammunition, communications equipment, and other materiel needed for the effort, as well as paying the salaries of fighters and full-time workers in the organization. Often the underground component of a movement plays a leading role in securing funds for military units to pay salaries and buy supplies. In the Philippines, it was a prime responsibility of the communist Politburo in Manila to obtain money for the Hukbalahap movement; and in Malaya, the Min Yuen was the major supplier of money to the rebels, obtaining many funds by extorting money from large landowners and transportation companies and by appropriating cash from communist-dominated unions.

Funds are also needed for other purposes, such as paying bribes. Insurgencies thrive under corrupt governments, and undergrounds often disperse money to key officials to obtain their protection or silence. Bribery also plays a part in subversion and the gathering of intelligence. Additionally, some insurgencies require funds to support their social outreach work and shadow government activities. Just as legitimate state governments struggle with the rising cost of medical care, unemployment insurance, food aid, housing subsidies, and pensions, some insurgent movements also struggle to provide similar services in an attempt to build legitimacy and undermine ruling authorities. The aid often serves a dual function by providing care for key constituencies and by offering cover for illegal and violent activities. However, providing various social welfare benefits is expensive and requires a sustained and reliable source of income.

Insurgent groups typically obtain resources from a variety of activities, including kidnapping, extortion, and taxation, as well as through investments, trade, and diaspora remittances.[154] Groups may also obtain military equipment and financial resources through foreign governments, as well as through selling various "lootable" resources, as was the case in Sierra Leone (diamonds), Burma (timber), the Democratic Republic of Congo (gold), and Côte d'Ivoire (cocoa). The support and resources can be immensely valuable to resistance groups, enabling them to carry out activities that otherwise would not have been possible. For instance, Nigerian militants are able to trade oil for guns in the Niger Delta;[155] Guinea provided logistical support to enable the Liberians United for Reconciliation and Democracy (LURD) to move military supplies to operational zones;[156] and various African groups have benefited from the provision of safe havens in neighboring countries. Examples include Burundi and Rwandan rebels in the Democratic Republic of the Congo; Congolese rebels in Rwanda and Uganda; the Sudan People's Liberation Army in Ethiopia, Eritrea, and Uganda; and Ugandan rebels in the Congo and Sudan.[157]

The type of resources and assistance available to groups may impose various constraints and demands that may not be immediately apparent. For a group to be able to profit from the presence of natural resources, it first must control the territory in which the resources are located. Such a requirement imposes obvious demands on tactics and operations. Sometimes rebel control of natural resources fluctuates over time. Rebels in Sierra Leone maintained control over diamonds only after 1997, whereas previously it had only been able to maintain control for several months at a time and had come close to defeat (in 1992 and 1995).[158] Additionally, a group must be able to harvest natural resources, which requires either compelling a local population to extract the resources or doing so themselves by diverting personnel away from resistance activities. Both options are costly, particularly because the labor-intensive nature of resource extraction may require the diversion of combatants from their military duties.[159]

Lastly, contextual factors and a group's goals and tactics impact its sourcing of various resources. Geography may play a role, as was the case in the 1990s with the Bosnian government, which experienced difficulty accessing international arms markets because it was landlocked and surrounded by neighbors.[160] If a resistance group faces a weak government with a poorly resourced military unable to project power over long distances, it may not need to develop expensive capabilities to resist state forces.[161] Furthermore, if a group opts for irregular tactics and does not desire to hold territory, then it does not need to develop a costly conventional army.

However, for groups intending to hold territory and battle a conventional army, resourcing presents more challenges. The Zimbabwe African People's Union (ZAPU), with a desire to hold territory in Rhodesia and replace the white-minority regime, combined with a reliance on conventional tactics (in addition to guerrilla warfare) imposed strong requirements on its resourcing capabilities. In this case, the group benefited from the existence of an external sponsor, the Soviet Union, which provided various conventional capabilities (tanks, planes, etc.) and associated training to assist the group in taking on the white-minority government in Salisbury, now Harare.

Resourcing Endowments and Civilian Targeting

Insurgent groups, based on environmental conditions, can have different resource endowments that impact the strategies and tactics the group adopts later. As discussed throughout the previous chapters, insurgent groups face significant challenges in building an organization that is capable of neutralizing a conventional state army. Jeremy Weinstein theorized that groups use different sorts of resource endowments to overcome these barriers. Some groups have economic endowments that come from natural resource extraction, taxation, criminal activity, or external sponsorship. By contrast, other groups have social endowments, which include shared beliefs, expectations, and norms within relevant religious or ethnic groups.[162]

Weinstein argues that the type of endowment with which an insurgent leader begins shapes the organization in numerous ways. One of the most important is the membership profile of the group. Groups with economic endowments attract "low-commitment" investment recruits primarily motivated by financial gain because leaders of these groups have enough resources to offer attractive selective incentives.[163] Groups operating only with social endowments tend to attract "high-commitment" recruits that are more interested in activist rebellion than short-term economic gain.

The most important effect of these different endowments and membership profiles is the impact they have on how the insurgents use violence. Groups with economic endowments are more likely to engage in indiscriminate violence that disproportionately targets civilians. The lure of short-term economic gain tends to win out over pursuit of long-term gains more beneficial to the local community. On the other hand, groups with social endowments exhibit greater restraint and discipline when using violence, tending to establish a more collaborative relationship with host populations. Because these groups lack resources, they must rely more on a collaborative relationship with the community. This means that leaders of these groups are more likely to consider the needs of the community and develop mechanisms to share power that act as a check on the leader's actions.[164] The endowment of a group is decided fairly early in a group's organizational life. This means that early developments have a significant impact on the decisions and behaviors of the group throughout its history, constraining the decisions and behavior of insurgent leaders.[165, 166]

The differing membership profiles of insurgent groups with economic and social endowments mean that the groups have different relationships with host populations. Groups with an economic endowment have less need to seek the consent or collaboration of the civilian population to raise resources and thus tend to engage in more looting.[167] Additionally, recruits of these groups tend not to have ties with the local population, which makes identifying defectors difficult.[168] Groups that receive external support similarly have fewer incentives to adopt a "hearts and minds" strategy, but this effect can be diminished when the external supporter, or principal, has strong commitments to human rights protections that help to mitigate wartime atrocities by the resistance, the agent.[169]

In contrast, groups with social endowments cannot entice potential recruits with opportunities for looting. In particular, they must attract recruits that maintain ethnic, religious, or ideological ties with a local population. Insurgents make appeals to this shared affiliation with aggrieved groups, promising that a rebel victory will translate into collective benefits for the group.[170] These groups typically obtain resources by reaching an agreement with the host population and can maintain internal discipline by relying on group norms and guidelines governing how combatants should behave. Under such conditions, groups have a greater capacity to selectively wield violence and discipline the use of force.[171]

Examples of groups that were organized along social endowments are the NRA of Uganda and the national-level organization of the Shining Path in Peru. The NRA was organized on ethnic lines, while the Shining Path organized along ideological lines, and both attracted committed recruits, used violence selectively, and established structures that encouraged cooperation and discipline.[172] In contrast, the Mozambican National Resistance (Resistencia Nacional Moçambicana, or RENAMO) and a regional committee of the Shining Path in the Upper Huallaga Valley were organized along economic endowments; RENAMO was sponsored by an external patron, and the Upper Huallaga Valley group financed its operations through the drug trade. These two groups attracted short-term opportunists, lacked procedures for disciplining behavior, and committed widespread atrocities against noncombatants.[173]

CONCLUSION

Leaders of insurgent groups make numerous strategic, organizational, and resourcing decisions that shape how the armed group operates. One of the most impactful decisions regards whether the group uses predominantly violent or predominantly nonviolent tactics. Many full-blown civil wars and other violent conflicts began as nonviolent opposition movements; therefore, explaining why the transition occurs is a research imperative. The ARIS project explicitly regards groups that rely predominantly on nonviolent tactics as taking part in resistance; these groups simply use different tactics than groups that rely on violence. Conventional wisdom holds that groups adopt violent tactics for instrumental reasons related to the effectiveness of violence in achieving political objectives.

Another explanation for the transition to violence discussed in this chapter sees violence emerging from repeated interactions between actors party to a conflict. The arenas of interaction include those between the movement and its political environment, among movement actors, between the movement and security forces, and between a movement and a countermovement. The transition to violence is triggered by mechanisms related to competition during intense periods of mobilization. Opposition movements are expected to compete with their targets, usually the state, in pressing for their political claims. However, competition also occurs among movements as individuals and smaller groups within it vie for power and influence. Although resistance movements are often treated as monolithic actors, most are heterogeneous, made up of individuals and smaller groups with different views on ideology and tactics. This internal competition for power is especially likely when the movement faces repression from security forces, generating splinter groups that have divergent opinions about how to respond. Similarly, competitive escalation explains how repeated interactions between activists and the police can contribute to violence as each move and countermove raises the stakes of the game.

Overcoming the significant barriers faced by resistance movements targeting powerful states requires leaders to make decisions regarding organizational structure and resourcing. While seemingly prosaic, the decisions have far-ranging repercussions that impact how violence is deployed. Organizational structures have been found to impact the type and scale of violence. Hierarchical organizations are able to sustain a higher intensity and duration of violence than networked or compartmentalized structures. Similarly, organizations structured according to function are more likely to use indiscriminate violence to pursue their goals. Networked or compartmentalized structures, because they lack robust command and control functions, face unique dilemmas. Without the ability to enforce strategic discipline on the group, leaders have difficulty enforcing rules, including those against the use of indiscriminate violence, which can compromise a group's wider goals and support from the host population.

Most groups require some safe space from which to plan and operate. Although territorial control is an important strategic necessity, not all groups achieve it. In the control-collaboration model, the extent of territorial control is correlated with the use of indiscriminate or selective violence. The more territorial control a group enjoys, the more it is able to rely on selective violence to consolidate its gains and assert control over the population. Selective violence is preferable to indiscriminate violence insomuch as it enables greater support from the host population. Indiscriminate violence is more likely to be practiced by armed actors in areas of contested control. Furthermore, once they establish territorial control, armed actors are able to engage in shadow governance activities, increasing their legitimacy by mimicking the functions of a sovereign state.

Resources are a critical enabler in an armed group's struggle against a more powerful state. Armed groups rely on a variety of revenue streams, including taxation, extortion and kidnapping, natural resource looting, and external state support. A group's endowments, whether economic or social, influence its membership profile. The economic endowments afforded by lootable natural resources or external sponsorship attract low-commitment investors interested primarily in short-term gain. On the other hand, a group with social endowments based on shared ideologies or norms attracts high-commitment investors that are interested in the long-term political goals of the group. Low-commitment membership encourages poorer relations with the host population, leading to more predatory behavior and indiscriminate violence. The activists associated with social endowments, however, rely more heavily on the host population for support, forging greater links with locals who provide more intelligence and resources. As a result, groups with social endowments are more likely to rely on selective violence and exhibit greater disciplinary control over their members.

ENDNOTES

1   TC 18-01, *Special Forces Unconventional Warfare* (Washington, DC: Headquarters, Department of the Army, 2010).

2   Political scientists' preference for the term *civil war*, as opposed to *insurgency*, stems from a focus on the conflict as the unit of analysis, rather than the individual organizational actors party to the conflict. In research related to defense and security, the prevalent use of the term *insurgency* to describe similar observable phenomena emerges from the preference to focus on individual actors, usually insurgent groups waging an armed campaign against an incumbent regime. While the former approach offers a great deal of valuable information on aggregate patterns in conflict, particularly when combined with large-n research, the latter yields rich detail about organizational characteristics, practices, and internal dynamics that transition well to education and training on

optimal execution of unconventional warfare, foreign internal defense, and counter-terrorism missions. Regardless, while the terms *civil war* and *insurgency* are not always synonymous, they are most often used to describe the same conflict processes, whether those in Colombia, Syria, or Iraq.

3 Will H. Moore, "Action-Reaction or Rational Expectations? Reciprocity and the Domes-tic-International Conflict Nexus during the 'Rhodesia Problem,'" *Journal of Conflict Resolution* 39, no. 1 (1995): 129–167.

4 Jacob N. Shapiro, *The Terrorist's Dilemma: Managing Violent Covert Organizations* (Princeton: Princeton University Press, 2013), 3–10.

5 Erica Chenoweth and Maria J. Stephan, *Why Civil Resistance Works: The Strategic Logic of Nonviolent Conflict* (New York: Columbia University Press, 2012), 12.

6 For a more extensive list of nonviolent tactics, see Table 10-1 in Bos, ed., *Human Factors Considerations of Undergrounds and Insurgencies*.

7 Ibid., 12; Nathan Bos, ed., *Human Factors Considerations of Undergrounds and Insurgencies* (Fort Bragg, NC: United States Special Operations Command, 2012), 280.

8 Jon Kleinberg, "The Convergence of Social and Technological Networks," *Communications of the ACM* 53, no. 11 (2008): 66–72.

9 Lorenzo Bosi, Chares Demetriou, and Stefan Malthaner, eds, "A Contentious Politics Approach to the Explanation of Radicalization," in *Dynamics of Political Violence: A Process Oriented Perspective on Radicalization and the Escalation of Political Conflict*, (Surrey, England: Ashgate, 2014), Chapter 1.

10 Ibid., 5.

11 This discussion on the threshold of violence is adapted from the section "Managing the Threshold of Violence" in Pinczuk, ed., *ARIS Threshold of Violence*.

12 For a discussion of the LTTE's transition from irregular to conventional tactics, please see Nix and Marshall, "Liberation Tigers of Tamil Eelam (LTTE)," 194–197.

13 Maegen Nix and Shana Marshall, "Liberation Tigers of Tamil Eelam (LTTE)," in *ARIS Casebook on Insurgency and Revolutionary Warfare Volume II: 1962–2009*, ed. Chuck Crossett (Fort Bragg, NC: USASOC, 2012), 194–197.

14 In this literature, the concept of radicalization is used to describe a group's transition from a predominantly nonviolent group to a predominantly violent one. Alimi, Bosi, and Demetriou define it as "the development of extreme ideology and/or the development of violent forms of contention, including categorical indiscriminate violence (or terrorism) by a challenging group."

15 Eitan Y. Alimi, Lorenzo Bosi, and Chares Demetriou, "Relational Dynamics and Processes of Radicalization: A Comparative Framework," *Mobilization: An International Journal* 17, no. 1(2012): 7–26.

16 Various macrostructural factors are discussed in chapter 2.

17 Gianluca De Fazio, "Intramovement Competition and Political Outbidding as Mechanisms of Radicalization in Northern Ireland, 1968–1969," in *Dynamics of Political Violence: A Process Oriented Perspective on Radicalization and the Escalation of Political Conflict*, eds. Charles Demetriou et. al, (Surrey, England: Ashgate, 2014), 115.

18 Alimi, Bosi, and Demetriou, "Relational Dynamics and Processes of Radicalization," 7–26.

19 Donatella della Porta, "Competitive Escalation During Protest Cycles: Comparing Left-Wing and Religious Conflicts," in *Dynamics of Political Violence: A Process Oriented Perspective*

on Radicalization and the Escalation of Political Conflict, eds. Demetriou et al., (Surrey, England: Ashgate, 2014), 93.

[20] Sidney Tarrow, *Power in Movement: Social Movements and Contentious Politics* 3rd ed. (Cambridge and New York: Cambridge University Press, 1998), 142.

[21] The transference of protest techniques and tactics is called diffusion. Diffusion occurs horizontally across different protest units from a conscious decision to import an innovation based on its success in its place of origin or as a result of collaboration among protest units connected through transnational networks.

[22] David Patel, Valerie Bunce, and Sharon Wolchik, "Diffusion and Demonstration," in *The Arab Uprisings Explained: New Contentious Politics in the Middle East*, ed. Marc Lynch (New York: Columbia University Press, 2014), 58–59.

[23] della Porta, "Competitive Escalation During Protest Cycles," 94–95; Tarrow, *Power in Movement*, 141–160.

[24] Charles Tilly, *Popular Contention in Great Britain, 1758–1864* (Cambridge, MA: Harvard University Press, 1995), 41.

[25] D. A. Binchy, "A Pre-Christian Surivial in Mediaeval Irish Hagiography," in *Ireland in Medieval Europe: Studies in Memory of Kathleen Hughes*, ed. Dorothy Whitelock (Cambridge: Cambridge University Press, 1982), 168.

[26] For an analysis of how the PIRA successfully used hunger strikes against the British, see Crossett and Newton, "The Provisional Irish Republican Army (PIRA)," 316–319.

[27] Chuck Crossett and Summer Newton, "The Provisional Irish Republican Army (PIRA): 1969-2001," in *Casebook on Insurgency and Revolutionary Warfare Volume II: 1962–2009*, ed. Chuck Crossett (Fort Bragg, NC: USASOC, 2012), 315–317.

[28] Charles Tilly, *Regimes and Repertoires*, (Chicago: University of Chicago, 2010), 41.

[29] Tarrow, *Power in Movement*, 142.

[30] Ibid., 29–42.

[31] Many of the familiar forms of nonviolent protest used by contemporary resistance movements have their roots in nineteenth-century France and England. In the tumultuous century following the French Revolution, angry, urban poor in Paris first began to use the barricade and urban insurrection to press their claims against the corrupt regime of the French ruling class. Similarly, in England, mass petitions first emerged around this time as organized nonstate actors brought petitions to the English parliament, protesting everything from taxes to slavery.

[32] della Porta defines political violence as "the use of physical force to damage a political adversary."

[33] della Porta, *Clandestine Political Violence*, 6.

[34] Tarrow, *Power in Movement*, 94–95.

[35] Ibid., 94–95.

[36] Ibid., 96–98.

[37] Ibid., 98–101.

[38] More recently, developments in information technology, particularly social media, have had a profound impact on how some resistance movements emerge, mobilize, operate, fail, and succeed. These issues are discussed in a separate forthcoming edited volume, *ARIS Resistance in the Cyber Domain*.

[39] Sabine C. Carey, "The Dynamic Relationship Between Protest and Repression," *Political Research Quarterly* 59, no. 1 (2006): 1-11.

[40] Christian Davenport and Will H. Moore, "The Arab Spring, Winter, and Back Again?(Re) Introducing the Dissent-Repression Nexus with a Twist," *International Interactions* 38, no. 5 (2012): 704-713.

[41] della Porta observes that it is misleading to describe all policing efforts as repression. She uses the term *policing of protest*, which she simply defines as "the police handling of protest events." Her definition captures violent, repressive measures alongside routine policing aimed at restoring law and order.

[42] Donatella della Porta, *Social Movements, Political Violence, and the State: A Comparative Analysis of Italy and Germany* (Cambridge: Cambridge University Press, 2006), 55–56.

[43] See chapter 3 for a discussion of how different types of regimes impact the ability of a reistance movement to mobilize its supporters.

[44] Alimi, et al., "Relational Dynamics and Processes of Radicalization."

[45] Tarrow, *Power in Movement*, 95–96; Donatella della Porta, *Social Movements, Political Violence, and the State: A Comparative Analysis of Italy and Germany* (Cambridge: Cambridge University Press, 2006), 55–82.

[46] Doug McAdam, *Political Process and the Development of the Black Insurgency, 1930–1970* (Chicago: University of Chicago, 1982).

[47] The use of the term *insurgent* to describe the black movement is Doug McAdam's, not the decision of the ARIS authors.

[48] Doug McAdam, "Tactical Innovation and the Pace of Insurgency," *American Sociological Review* 48, no. 6 (1983): 735–754.

[49] della Porta, "Competitive Escalation During Protest Cycles," 95.

[50] Alimi et al., "Relational Dynamics and Processes of Radicalization."

[51] della Porta, "Competitive Escalation During Protest Cycles," 98.

[52] Tarrow, *Power in Movement*, 207.

[53] Ibid., 96.

[54] della Porta, "Competitive Escalation During Protest Cycles," 98–103.

[55] Eitan Alimi, Lorenzo Bosi, and Chares Demetriou, "Relational Dynamics and Processes of Radicalization," 7–26.

[56] de Fazio, "Intramovement Competition and Political Outbidding," 117–120.

[57] Ibid., 124–131.

[58] For a detailed case study of the PIRA, see Crossett and Newton, "The Provisional Irish Republican Army (PIRA)."

[59] Chuck Crossett and Summer Newton, "The Provisional Irish Republican Army (PIRA): 1969-2001," in *Casebook on Insurgency and Revolutionary Warfare Volume II: 1962–2009*, ed. Chuck Crossett (Fort Bragg, NC: United States Army Special Operations Command, 2012), 379–322

[60] de Fazio, "Intramovement Competition and Political Outbidding," 478.

[61] This section is adapted, revised, and updated from Bos, ed., *Human Factors Considerations of Undergrounds in Insurgencies*, Chapter 3.

[62] In terms of phasing, the insurgent groups studied in this research are generally in the institutionalization state of resistance.

[63] W. Sam Lauber, ed., *ARIS Understanding the States of Resistance* (Fort Bragg, NC: United States Army Special Operations Command, forthcoming publication).

[64] Portions of this section are adapted, revised, and updated from Bos, *Human Factors Considerations of Undergrounds in Insurgencies*, Chapter 3.

[65] David S. Alberts and Richard E. Hayes, *Power to the Edge* (CCRP Publication Series, 2003).

[66] Leaders, however, are constrained in decisions regarding organizational structure as insurgent groups are mobilized based on pre-existing social networks with their own organizational configurations, including political parties, kinship groups, student groups, or religious associations.

[67] Paul Staniland, *Networks of Rebellion: Explaining Insurgent Cohesion and Collapse* (Ithaca, NY: Cornell University Press, 2014), 8–9.

[68] Lindsay Heger, Danielle Jung, and Wendy H. Wong, "Organizing for Resistance: How Group Structure Impacts the Character of Violence," *Terrorism and Political Violence* 24, no. 5 (November 2012): 743–768.

[69] Ibid., 745.

[70] Ibid., 757–763.

[71] Ibid., 748.

[72] Ibid.

[73] Ibid., 749.

[74] Jacob N. Shapiro, *The Terrorist's Dilemma: Managing Violent Covert Organizations* (Princeton: Princeton University Press, 2013).

[75] Ibid., 34–35.

[76] Ibid., 38.

[77] Ibid., 46–47.

[78] Ibid., 49.

[79] Ibid., 50.

[80] Maryjane Osa, "Networks in Opposition: Linking Organizations through Activists in the Polish People's Republic," in Mario Diani and Doug McAdam (eds.), *Social Movements and Networks: Relational Approaches to Collective Action* (Oxford University Press, Oxford: 2003).

[81] This section includes research that pertains mostly to nonviolent movements, not violent groups operating in the institutionalized state phase of resistance.

[82] Timothy Garton Ash, The Polish Revolution: Solidarity (New Haven: Yale University Press, 1999), 24.

[83] Approximately one-third of the Polish population joined independent, professional, social, or political organizations from 1976 to 1981. After the communist government declared martial law in 1981, one in five Poles participated in at least one collective protest, an astonishing figure. See Crossett and Newton, " Solidarity," 645–672, for a comprehensive analysis of the Solidarity resistance movement.

[84] Chuck Crossett and Summer Newton, "Solidarity," in *Casebook on Insurgency and Revolutionary Warfare Volume II: 1962–2009*, ed. Chuck Crossett (Fort Bragg, NC: United States Army Special Operations Command, 2013), 645–672

[85] Osa, "Networks in Opposition," 98.

[86] Another important feature of the resistance movement was that it brought together an ideologically diverse collection of groups, including Catholics, nationalists, secular intellectuals, and labor activists subsumed under a broader movement that proffered master frames or themes that all groups could rally around, including human rights, national culture, and social autonomy from the state.

[87] Osa, "Networks in Opposition," 101.

[88] Unless otherwise noted, this section on framing is based on the discussion in Snow, "Identity Dilemmas, Discursive Fields, Identity Work, and Mobilization," 267–268.

[89] David Snow, "Identity Dilemmas, Discursive Fields, Identity Work, and Mobilization: Clarifying the Identity-Movement Nexus," in *Dynamics, Mechanisms, and Process: The Future of Social Movement Research*, eds. Jacquelien von Stekelenburg, Conny Roggeband, and Bert Klandermans (Minneapolis: University of Minnesota Press, 2013), 267–268.

[90] Snow, "Identity Dilemmas, Discursive Fields, and Identity Work," 273.

[91] This discussion of framing is adapted from Agan, ed., *ARIS Narratives and Competing Messages*, 35–43.

[92] Summer D. Agan, ed., *ARIS Narratives and Competing Messages* (Fort Bragg, NC: United States Special Operations Command, forthcoming publication), 35–43

[93] Erving Goffman, *Frame Analysis: An Essay on the Organization of Experience* (New York: Harper Colophon, 1974), 21.

[94] Robert D. Benford and David A. Snow, "Framing Processes and Social Movements: An Overview and Assessment," *Annual Review of Sociology* 26 (2000): 614.

[95] Ibid., 616–7.

[96] David A. Snow and Robert D. Benford, "Ideology, Frame Resonance, and Participant Mobilization," *International Social Movement Research* 1, no. 1 (1988): 198.

[97] Benford and Snow, "*Framing Processes and Social Movements*," 617.

[98] Ibid.

[99] Jerry Conley, "Movement for the Emancipation of the Niger Delta (MEND)," in *Casebook on Insurgency and Revolutionary Warfare Volume II: 1962–2009*, ed. Chuck Crossett (Ft. Bragg, NC: United States Army Special Operations Command, 2012), 738–9, 743, 748.

[100] Sanaz Mirzaei, "Palestinian Liberation Organization (PLO): 1964–2009," in *Casebook on Insurgency and Revolutionary Warfare Volume II: 1962–2009*, ed. Chuck Crossett (Ft. Bragg, NC: United States Army Special Operations Command, 2012), 286, 295, 299, 302–3.

[101] Benford and Snow, "*Framing Processes and Social Movements*," 617.

[102] Assaf Moghadam, "Mayhem, Myths, and Martyrdom: The Shi'a Conception of Jihad," *Terrorism and Political Violence* 19, no. 1 (2007): 133.

[103] Benford and Snow, "Framing Processes and Social Movements," 623.

[104] Ron Buikema and Matt Burger, "Sendoro Luminoso (Shining Path)," in *ARIS Casebook on Insurgency and Revolutionary Warfare Volume II: 1962–2009*, ed. Chuck Crossett (Fort Bragg, NC: United States Army Special Operations Command, 2012), 75, 79–82, 94–6.

[105] Haggay Ram, *Myth and Mobilization in Revolutionary Iran: The Use of the Friday Congregational Sermon* (Washington, DC: American University Press, 1994), 37.

[106] Ibid.

[107] Robert Ardrey, *The Territorial Imperative: A Personal Inquiry into the Animal Origins of Property and Nations* (New York: Atheneum, 1966).

[108] Monica Duffy Toft, "Territory and War," *Journal of Peace Research* 51, no. 2 (2014), 186.

[109] Ibid., 187.

[110] J. Bowyer Bell, "Dragonworld (II): Deception, Tradecraft, and the Provisional IRA," *International Journal of Intelligence and Counterintelligence* 8, no. 1 (1995): 21–50.

[111] In the months leading up to the 1969 split between the official IRA and the PIRA in Northern Ireland, riots between Catholics, Protestants, and the Northern Irish police force led to the establishment of "no-go" areas in Derry. Participants in the Bogside riots, as they were called, established the small islands of liberated territory to plan the defense of the beleaguered Catholics. The barricades were not removed until the British Army arrived and dismantled them in Operation Motorman.

[112] Chuck Crossett and Summer Newton, "The Provisional Irish Republican Army," 302–303.

[113] Territorial control is more likely to occur in states with weak governments and poor economic development.

[114] Luis de la Calle and Ignacio Sánchez-Cuenca, "Rebels without a Territory: An Analysis of Nonterrirotial Conflicts in the World, 1970-1997," *Journal of Conflict Resolution* 56, no. 4 (2012): 580–603.

[115] Luis de la Calle and Ignacio Sánchez-Cuenca, "How Armed Groups Fight: Territorial Control and Violent Tactics," *Studies in Conflict & Terrorism* 38, Vol. 10, 796–797.

[116] Ibid., 797.

[117] The authors also considered the possibility of reverse causality. That is, they explored whether the type of tactics a group uses impacts its territorial control. They largely dismissed this as a possibility, arguing that in most cases, groups decide from the beginning to act underground without aspiring to control territory or they very quickly gain territory and function as a guerrilla army.

[118] Ibid., 798–800, 805.

[119] de la Calle and Sánchez-Cuenca, "How Armed Groups Fight," 808–809.

[120] Luis de la Calle and Igancio Sánchez-Cuenca, "In Search of the Core of Terrorism," *International Studies Review* 14 (2012): 475–497.

[121] Stathis Kalyvas's control-collaboration model can be applied to numerous issues related to the dynamics of violence. In chapter 3, the same model is used to explain mobilization in violent conflicts. See Kalyvas, *The Logic of Violence in Civil War,* for a good description of the control-collaboration model. See also Kalyvas, "Micro-Level Studies of Violence in Civil War," for an abbreviated explanation of the same model.

[122] Stathis Kalyvas, *The Logic of Violence in Civil War* (New York: Cambridge University Press, 2006), 195–209; Stathis Kalyvas, "Micro-level Studies of Violence in Civil War: Refining and Extending to Control-Collaboration Model," *Terrorism and Political Violence* 24, no. 4(2012): 658–668.

[123] Stathis Kalyvas, *The Logic of Violence in Civil War* (New York: Cambridge University Press, 2006), 195–209; Stathis Kalyvas, "Micro-level Studies of Violence in Civil War: Refining and Extending to Control-Collaboration Model," *Terrorism and Political Violence* 24, no. 4(2012): 658–668.

[124] Ibid.

[125] Ibid.

[126] Ibid.

[127] Ibid., 149.

[128] Kalyvas's research focused on the Argolid region of southern Greece.

[129] D.M. Condit, *Case Study in Guerrilla War: Greece during World War II,* ed. Erin M. Richardson (Fort Bragg, NC: United States Army Special Operations Command, 2014); Kalyvas, *The Logic of Violence in Civil War,* 266–298.

[130] Portions of this section are adapted, revised, and updated from Newton and Leonhard, "Shadow Government," 131–168.

[131] Summer Newton and Robert Leonhard, "Shadow Government," in *Undergrounds in Insurgent, Revolutionary, and Resistance Warfare*, ed. Robert Leonhard (Fort Bragg, NC: United States Army Special Operations Command, 2013), 131–168.

[132] David Howell Petraeus and James F. Amos, *FM 3-24 2006 Counterinsurgency*, (Boulder, CO: Paladin Press), 1–65.

[133] Nelson Kasfir, "Guerrillas and Civilian Participation: The National Resistance Army in Uganda, 1981–86," *Journal of Modern African Studies* 43, no. 2 (2005): 272.

[134] Ibid.

[135] William Maley, "Building Legitimacy in Post-Taliban Afghanistan," in S*tate Building, Security and Social Change in Afghanistan* (San Francisco: Asia Foundation, 2008): 12, http://asiafoundation.org/resources/pdfs/2008surveycompanionvolumefinal.pdf.

[136] Ibid, 13–15.

[137] Guillermo Pinczuk, *Case Studies in Insurgency and Revolutionary Warfare – Sri Lanka (1976–2009)* (Fort Bragg, NC: United States Army Special Operations Command, 2013).

[138] Nelson Kasfir, "Guerrillas and Civilian Participation: The National Resistance Army in Uganda, 1981–86," *Journal of Modern African Studies* 43, no. 2 (2005): 286.

[139] James B. Love, *Hezbollah: Social Services as a Source of Power* (Hurlbert Field, FL: JSOU Press, 2010), 21.

[140] Ibid., 26.

[141] Ibid.

[142] Tony Addison and Syed Mansoob Murshed, *The Fiscal Dimensions of Conflict and Reconstruction* (Helsinki: United Nations University, World Institute for Development Economics Research, 2001), 1–2.

[143] Ibid., 5.

[144] However, some insurgencies still benefit from the generous support of external supporters.

[145] Paul Kingston and Ian S. Spears, eds., *States-within-States: Incipient Political Entities in the Post-Cold War Era* (Basingstoke, UK: Palgrave MacMillan), 25.

[146] Zachariah Cherian Mampilly, *Rebel Rulers: Insurgent Governance and Civilian Life During War* (Ithaca: Cornell University Press, 2011), 123.

[147] This section has been adapted from the Guillermo Pinczuk, *Case Studies in Insurgency and Revolutionary Warfare – Sri Lanka (1976–2009)* (Fort Bragg, NC: United States Army Special Operations Command, 2013).

[148] Mampilly, Rebel Ruler, 94–95.

[149] Ibid., 111.

[150] Ibid., 112, 115.

[151] Ibid., 114.

[152] Ibid., 115–116, fn. 21.

[153] Portions of this section are adapted, revised and updated from Leonhard, ed., A*RIS Undergrounds in Insurgent, Revolutionary, and Resistance Warfare*, Chapter 4.

[154] Jennifer M. Hazen, *What Rebels Want: Resources and Supply Networks in Wartime* (Ithaca: Cornell University Press, 2013), 2.

[155] Ibid., 14.

[156] Ibid., 58.

[157] Ibid., 57.

[158] Ibid., 22.

[159] Ibid., 11–12.

[160] Ibid., 12.

[161] Ibid., 8–9.

[162] Jeremy M. Weinstein, *Inside Rebellion: The Politics of Insurgent* Violence (Cambridge University Press, Cambridge, 2007), 6–7.

[163] For a discussion of selective incentive in mobilization processes, see chapter 3.

[164] Weinstein, Inside Rebellion, 6–7.

[165] This is called path dependence, or when a set of circumstances or decisions intiates a self-reinforcing lock-in that limits and constrains changes in future behaviors. Path dependence means that "current and future states, actions, or decisions depend on the path of previous states, actions, or decisions." Path dependence raises questions about the relationship between structure and agency in the explanation of insurgent behavior. For a treatment of this fundamental methodological distinction and how it impacts research, see chapter 2. Notice also that Weinstein's theory of resource endowments and indiscriminate violence offers an alternative explanation to Kalyvas's control-collaboration model, which uses territorial control to explain patterns of selective and indiscriminate violence.

[166] Scott E. Page, "Path Dependence," *Quarterly Journal of Political Science* 1, no. 1 (2006): 87–115.

[167] Weinstein, *Inside Rebellion*, 9–10.

[168] Ibid., 10.

[169] Idean Salehyan David Siroky, and Reed M. Wood, "External Rebel Sponsorship and Civilian Abuse: A Principal-Agent Analysis of Wartime Atrocities," *International Organization* 68, no. 3 (2014): 633–661.

[170] Weinstein, *Inside Rebellion* , 9.

[171] Ibid., 10.

[172] Ibid., 14.

[173] Ibid.

CHAPTER 5.
WHY DO CONFLICTS END? CONFLICT TERMINATION, RECIDIVISM, AND MOVEMENT DECLINE

Termination is the formal end of fighting in a conflict, though not always the final end of the conflict as a whole.[1] Seemingly resolved civil wars and other domestic conflicts often remain at risk of recidivism—when a previously terminated conflict reignites. Special Forces should be aware of the dynamics and factors at play in conflict termination and recidivism as they could define mission success in not only bringing about the end of a conflict in line with mission objectives but also in securing strategic gains on the long-term horizon of the postconflict environment. Finally, the dynamics of social movement decline offer insights on the growth, maturation, and decline of insurgent and resistance movements.

Studies on termination or demobilization are stratified into several categories, including research on civil wars and social movements, predominantly nonviolent ones. Although there is some cross-pollination between these two areas of study, more work is required to gain a fuller comprehension of termination and demobilization processes. The study of civil wars focuses on conflict terminations via negotiated settlements, military victories, cease-fires, or informal declines in violence. Most of the research is based on statistical regressions using large, aggregate datasets, such as the UCDP/PRIO, that include hundreds of cases across the globe and reach back decades into the past.[2] The research indicates that in the post-Cold War era, more civil wars ended through negotiated settlements, usually brokered by third parties, than military victories. This poses some difficult challenges because, while politically expedient for third parties brokering the deal, negotiated settlements are especially vulnerable to recurrences of civil war violence. The settlements are plagued with trust and commitment problems, although those issues can be mitigated when strong third parties enforce the terms of the agreement, which lowers rates of recidivism.

Outside of the large-n studies on civil war, social movement theory and the contentious politics program offer a rich conceptual spectrum to think about how resistance movements decline or demobilize. The theoretical richness of the research derives from its emphasis on qualitative methods involving case studies, which involves deep historical studies of a small number of resistance movements. The research demonstrates that resistance movements can decline through mostly positive processes, such as facilitation or institutionalization, or negative processes, such as co-optation and repression, that inhibit the capacity of the group to continue operations. Furthermore, the dynamics of interaction between a resistance movement and state authorities, as external pressure, generate or exacerbate ongoing internal dynamics of the movement. Social movement theory is helpful because it does not treat movements as monolithic actors working single-mindedly toward a set of static, prescribed goals but rather as a group comprising heterogeneous actors who do not always agree on the best strategies and tactics for reaching the group's political objectives. Together, the external and internal pressures can lead to competition and factionalization that can contribute to demobilization.

CIVIL WAR TERMINATION

Conflict termination is classified into three major categories: victory, peace agreement, or other outcomes including cease-fires and unresolved de-escalations. In a victory, one warring party has been defeated and/or eliminated by its opponent or otherwise succumbed to the opponent's power, such as through capitulation. Negotiated settlements, on the other hand, are mutual political arrangements that warring parties enter to explicitly regulate or resolve the source of grievances.[3] There are also nominal negotiated settlements, when the negotiations themselves are merely a formality in victory, as one side speaks from a position of weakness and makes disproportionate concessions, effectively losing the conflict.[4] Other less-decisive termination outcomes, such as cease-fire agreements and de-escalations of fighting below recognized thresholds, also occur.[5, 6] Finally, conflicts can often end "under unclear circumstances where fighting simply ceases."[7]

Negotiated settlements or peace agreements are mutually agreed upon political arrangements that warring parties enter into to explicitly regulate or resolve the source of grievances fueling the conflict.

Since the end of the Second World War, conflict termination has undergone significant changes concurrent with shifts in the international system. One of these key changes followed the end of the Cold War. Initially, researchers noted a surge in the number of civil wars, leading some to speculate anarchy would reign in the absence of bipolarity in the international system provided by the US–USSR rivalry.[8] The spike in the number of civil wars peaked in 1992, but many of those conflicts were contained by 1996.[9] In the same period, incidences of interstate wars plummeted, a dramatic change from past centuries that had witnessed near unceasing conflict between sovereign states. Currently, most armed conflicts are intrastate conflicts or conflicts that occur within a state rather than between states. In 2014, for instance, of the forty active conflicts (defined as those resulting in more than twenty-five battle deaths per year) worldwide, only one of those conflicts was between states (India and Pakistan), and it resulted in fewer than fifty fatalities. However, states continue to fight with one another through other means in so-called internationalized civil wars in which states provide external support to armed actors in the conflict.[10, 11]

The method by which conflicts terminate has also undergone significant systemic shifts. More and more conflicts are ending not through a decisive military victory by either the incumbent government or rebels but by negotiated settlements. Beginning in 1940 and until the end of the 1980s, most civil wars ended via a decisive military victory, whether by rebel forces or the incumbent government. From 1940 to 1999, civil wars were four times as likely to end via a military victory as a negotiated settlement.

However, beginning in 1990, not only did an astonishing number of civil wars (many of which had begun in the Cold War era) end, but many of them ended not through a military victory but by a negotiated settlement between the opposing sides. In the 1990s, only four out of every ten civil wars ended through a military victory. Of the thirty-seven civil wars that ended in the 1990s, fifteen ended in a military victory, fifteen by negotiated settlement and seven by a cease-fire/stalemate.[12] When the dataset researchers used to calculate those figures is extended to include low-intensity conflicts, the difference is even more striking. In the period from 1990 to 2005, of the 147 conflicts recorded, twenty-seven terminated through a negotiated settlement, while only twenty terminated via a military victory.[13]

The transformation of conflict termination in the post-Cold War era has increased opportunities for the international community to impact conflict outcomes, but it also revealed disturbing deficiencies in negotiated settlements. Conflicts terminated through such settlements are twice as likely to reignite as those that end in a clear military victory. The most stable outcomes, surprisingly, are rebel military victories. Moreover, in the aftermath of a rebel victory, democratization levels increase over time. The opposite, however, is true of government victories. Governments that have recently experienced a civil war, even if victorious, typically increase repression in the aftermath, wary of further openings of the political arena in light of recent political violence.[14]

The efficacy of nonviolent resistance campaigns in comparison with violent campaigns was discussed earlier. With some exceptions, most nonviolent resistance campaigns are more successful than violent ones. One of the reasons for the efficacy of such campaigns is the mass participation nonviolence enables. However, nonviolent campaigns are also important to consider because of what happens after the campaign succeeds. As will be discussed throughout this section, violent campaigns that lead to civil war, while they can be successful, often have enduring negative social outcomes for the states that experience them. Civil wars recur with depressing regularity, and it is difficult to consolidate democracy in the postconflict environment. Democracy is a more likely outcome when a nonviolent campaign, as opposed to a violent one, succeeds. Moreover, states are less likely to experience violent civil war after the resolution of a nonviolent campaign. Political conflicts may continue or arise, as in the aftermath of the Orange Revolution in Ukraine, but these subsequent conflicts are more likely to manifest as nonviolent. As a result, nonviolent campaigns generally produce better strategic outcomes in the long term.[15]

Rebel Victories

Rebel or insurgent victories have overall been less common than government victories, which account for more than twice as many outcomes, but the record for insurgent victories remains significant. In wars competing for control of the central government, rebel takeovers have succeeded roughly one in four times since 1955. Separatist insurgencies have historically been even more effective, almost half the time rendering opponents incapable of continuing the fight and gaining either de facto or significant regional autonomy provisions in 40 percent of cases.[16] If an insurgency succeeds, it is usually quick and decisive, pushing the government toward defeat at an accelerating rate. Insurgent defeats, however, tend to happen by degrees and at a gradually decelerating rate over time.[17]

Access to additional resources in neighboring countries significantly increases the likelihood of a rebel victory.[18] Bureaucratic effectiveness in the state has a detrimental effect on the potential for rebel victory, but democracy seems to have little impact on civil war outcomes. Protracted conflicts tend to favor insurgents, especially if military victory is not their primary goal. However, to benefit from a drawn-out conflict through organizational development or negotiated settlement, the rebels must first survive the government's initial counteroffensive early in the conflict. Other factors that make rebel success more likely in a protracted conflict include rough terrain, ineffective government, a nonethnic conflict, low levels of ethnic diversity, and a lack of United Nations (UN) intervention.[19]

Foreign interventions on behalf of rebel or insurgent forces have a statistically significant impact on the likelihood that the insurgent forces will win. While external support can act as a force multiplier, it can also lead to longer conflicts with the addition of another "veto player" at the bargaining table. External supporters might have an agenda that does not include conflict resolution; moreover, as they do not bear the direct cost of continued fighting, external supporters have fewer incentives to end the conflict.[20]

The favorable effect of external support is often because most third parties intervene on behalf of insurgencies only when the movements are most effective and at their strongest. Such interventions are most successful when conducted in support of groups that already possess relatively high levels of material capability and already pose a significant threat to the government. In addition to unconventional warfare, this also has important implications for counterinsurgency missions, as foreign third parties are most likely to intervene on behalf of the insurgents when they can have the greatest impact on the outcome of the conflict. Studies have shown that interventions on behalf of rebel movements are equally likely in both identity-based and ideological conflicts. However, higher numbers of shared borders increase the likelihood of such interventions, as neighboring territory often serves as staging grounds for military support.[21]

Government Victories

Government victories are more common than rebel victories in civil conflicts. In wars seeking central control since 1955, states achieved decisive military wins in 44 percent of cases, along with 13 percent where the government made concessions short of any power-sharing arrangements. Governments were able to render rebels incapable of continuing the fight in nearly half of separatist conflicts, but settlements through regional autonomy agreements are becoming more common; the frequency of this result grew from 20 percent of cases in the 1960s to 40 percent by 2008.[22] States generally defeat insurgencies gradually, winning by degrees as they push the rebels toward their breaking point at a decelerating rate (i.e., the closer an insurgency is to defeat, the slower any progress against them becomes). However, when states begin to lose to an insurgency, their declines tend to accelerate as they approach the breaking point, losing quickly and decisively.[23]

Factors favoring a government victory are less clear than for rebel victories. Large armies and effective bureaucracies have a detrimental effect on rebel victories but have only a minimal impact on the likelihood of a government victory. In fact, the use of a strong army against an insurgency can serve to exacerbate civil conflicts if the government does not quickly secure victory. Historical data suggest that the chances for a government victory drop over time if the government fails to secure victory near the outset. Because restraint from using the army can signal weakness and likewise encourage more resistance, governments trying to suppress resistance face difficult choices.[24]

Foreign interventions on behalf of a government do not significantly increase the likelihood of a government victory. This is partially because third parties most commonly intervene in civil wars when rebel groups pose the greatest threat to the government. This means progovernment interventions only appear ineffective because the majority involve the toughest cases. Third-party interventions to support at-risk governments are more likely in ideological conflicts than in identity-based conflicts (i.e., conflicts based on religious or ethnic divides). Unlike with proinsurgent interventions, the number of shared borders has no significant effect on the likelihood of third-party support for governments engaged in civil conflicts.[25]

Negotiated Settlement

Policy makers and researchers tend to focus on negotiated settlements as the most desirable of outcomes. The desire to reduce casualties and property damage, alongside the avoidance of losing sovereignty through occupation, drives the political desire for such agreements.

However, negotiated settlements also carry risk. Civil wars that end in these settlements are much more likely to reignite than those terminated through rebel or military victory. Generally, the armed actors in the conflict want the benefits of peace, but they also want to continue to hurt their enemies. They maintain a sense of legitimacy, a belief that they could have won, and an entitlement to more extensive claims. Together, the issues contribute to the risk for recurrence.[26]

Negotiated settlements are difficult to secure due to the trust and commitment problems inherent in the agreements, which typically include provisions for political power sharing, military integration, and mutual disarmament. Each side has limited means to reassure the other that they will abide by the agreement. These commitment and trust obstacles emerge especially in center-seeking civil wars, where the conflicts tend to be all-or-nothing, zero-sum games for control. Negotiated settlements are therefore more common in peripheral, autonomy-seeking wars.[27] State capacity does not seem to have any influence on the likelihood of a negotiated settlement, but UN intervention makes such settlements more likely.[28]

Competing Theories on the Problem of Civil War Resolution

There are competing theories among social science researchers regarding the primary challenges of terminating civil wars and maintaining a lasting peace after conflict resolution. The theories can be categorized into two general schools of thought: rationalist and ideational.

The rationalist school theorizes that adversaries in civil wars act based on the same cost-benefit calculations as interstate actors. Armed actors are thought to pursue continued armed conflict or resolution based on whether success is more likely through continued conflict or through a political settlement. However, there are challenges in domestic conflicts that can get in the way of negotiation and compromise. These problems may include (1) the uniquely high benefits of outright success in civil wars, either control of the state or independence; (2) the difficulty of dividing the benefits and stakes associated with domestic conflict; and/or (3) the problem of reaching mutually acceptable bargains in the context of ambitious leaders, extremist demands, poor communication, erratic outside aid, and fear.

The ideational school, on the other hand, sees internal conflicts as uniquely driven by intensely emotive conflicts based on values, as opposed to more rational bargaining, greed, or mere desire for dominance. These issues include ethnic and cultural identity, recognition, participation, and other deep and nonnegotiable value claims that make the prospect of common ground and compromise especially difficult.

Barbara F. Walter, "The Critical Barrier to Civil War Settlement," *International Organization* 51, no. 3 (1997): 341–343.

RECIDIVISM

After terminating, conflicts have a high probability of sliding back into conflict (i.e., recidivism or recurrence). Most countries that experience civil war will likely experience additional conflicts, a reality most scholars attribute to the structural conditions of the post-civil war environment that create incentives to resume violent conflict rather than accept the new status quo.[29] In other words, the problem of peace failure is both structure and agency based. Structurally, the type of conflict termination influences the likelihood of conflict recurrence (although these patterns are not simple).[30] In terms of agency, the details of peace agreements, power-sharing arrangements, governing coalitions, elections, peacekeeping operations, and other postconflict measures can have significant impacts on the decisions actors make in the near future.

Recidivism is a term used to describe the recurrence of civil war in a state that has experienced a period of peace or stability.

The nature and outcome of the previous civil war in a country has implications for the durability of postconflict peace.[31] Postconflict settlements are more likely to fail after high-intensity, high-casualty conflicts. The nature of the dispute underlying the conflict can also affect the likelihood of success. For example, politicoeconomic conflicts are less prone to recidivism than identity-based ones.[32] Conflicts with rebels seeking total control of the central government are also more likely to recur after termination than peripheral fights for independence or autonomy.[33] While there is debate among scholars regarding whether rebel or government victories are more likely to collapse into renewed conflict, most analysis suggests the former.[34] Specifically, rebel victories more often result in a durable peace than government victories, but only if the new regime can survive the initial few years. On the other hand, peace after a government victory is initially stable but often becomes more fragile over time if the government fails to address grievances by marshaling resources, institutional capacity, and the political will necessary for sustained reforms.[35]

While negotiated settlements have high rates of recidivism, some measures can positively affect the postconflict stability. Similar to rebel victories, the peace after a negotiated settlement is initially fragile but becomes more durable over time.[36] Several measures can be taken to make a negotiated peace more robust and likely to succeed. First, the presence of international peacekeeping operations in conjunction with a negotiated settlement makes the postconflict peace more durable than that following a government victory.[37] Second, multifaceted power-sharing arrangements that create extensive networks of institutions make former antagonists less likely to revert back to armed conflict. Rather than focusing solely on political decision-making, the more diversified and extensive the power-sharing institutions are (covering territorial, military, economic, and other dimensions of state power), the more likely agreement measures will be implemented and maintained.[38] Third, the presence of mediators in brokered negotiations that encourage more extensive power-sharing arrangements also improves the probability of long-term, self-enforcing peace.[39] Fourth, military power sharing and integration increase the likelihood of a long-lived peace, but only if implemented through a formally signed treaty, as opposed to merely ad hoc offers for integration from the government.[40] Finally, there is debate over the value of costly provisions in negotiated settlement and power sharing. Some argue that provisions with high costs make the stronger party more likely to either back out or seek renegotiation of the agreement, thereby destabilizing the arrangement and reigniting the conflict.[41] On the other hand, some argue that the implementation of provisions that require high degrees of sacrifice from both sides demonstrate the credibility of and increase confidence in the peace process.[42]

Elite interactions in governing coalitions and elections have implications on the success of power-sharing agreements. Elections may increase collaboration, but if they are rushed before the development of state institutions, they may instead serve to mobilize ethnonationalist competition or allow some elites to monopolize power, which may threaten to reignite conflict.[43] Studies suggest that recidivism is more likely if the governing coalition is small because those excluded from power have less to lose from resuming armed conflict. When included, both sides have institutional means to pursue political objectives that are less costly than armed violence. Large, inclusive governing coalitions are more likely in negotiated settlements and government victories than in rebel victories.[44] However, there are tradeoffs between building narrower and larger coalitions. Narrow coalitions are less likely to face internal defections but also increase the risk of civil war in the future. Large coalitions, while they decrease the likelihood of civil war recurrence, also increase the risk for coups and mutinies in the future.[45] Finally, if after elections, a regime seeks to dismantle democratic institutions and exclude opposition groups from future elections, the risk of recidivism increases.[46]

Research indicates that conflict recidivism is more common in the absence of international peacekeeping operations that provide security guarantees and transitional periods.[47] According to some authorities, peacekeeping operations reduce the risk of recurrence most significantly.[48] No matter the type of civil war, a country's most vulnerable moment is immediately after the end of fighting, but the longer a postconflict peace lasts, the less likely it is to break down into renewed conflict. For this reason, researchers believe that the international community needs to act quickly to begin peacekeeping operations to maximize the chance for success.[49] Studies suggest that to be effective, postconflict peacekeeping operations require a robust mandate, sufficient resources, and international commitment. The operations must also locally address the roots of hostilities while bolstering local capacities to address grievances. However, while peacekeeping activities are important to prevent recidivism, when the operations do not have the power or mandate necessary to end violence, the conflict environment necessitates multilateral enforcement operations or similar interventions.[50]

Other factors that scholars believe may influence conflict recidivism are the decentralization of power through regional autonomy or federalism, human rights trials, and territorial partition. Some argue that regional autonomy promotes confidence among groups in an agreement and provides the opposition with a regional base as a source of policy-making influence on the subnational level.[51] In the same way, federalism and the decentralization of power ensures that any loss of guaranteed seats for opposition figures at the national level would not be enough of a blow to their political influence to necessitate a return to conflict.[52] While the debate around the impact of human rights trials or tribunals on conflict recidivism is lively, preliminary analyses do not indicate that trials or tribunals significantly contribute to or detract from postconflict stability.[53] Others advocate territorial partition as the only way to end ethnic wars and prevent recurrence but caution that such measures could backfire, turning domestic conflicts into international war, thereby causing even greater instability and human suffering.[54, 55]

DECLINE IN SOCIAL MOVEMENTS AND NONVIOLENT RESISTANCE MOVEMENTS

Social movement theory and research on civil wars have established literature that focuses on the decline or demobilization phase of resistance movements. Research on demobilization of nonviolent resistance is less prolific than that on armed conflict termination. Nevertheless, the research on demobilization in social movement theory, because it deftly weaves features of structure and agency together, is an important contribution to understanding how resistance movements terminate.

One of the advantages of social movement theory is its rich conceptual spectrum of the numerous ways that resistance movements can cease to operate. The family of processes is referred to as "decline" in the social movement literature.[56] The Central Intelligence Agency's *Guide to the Analysis of Insurgency* outlines several of these outcomes as potential outcomes in the resolution phase of insurgencies.[57] In its *Understanding States of Resistance* research, ARIS researchers adopted resolution as the final phase of resistance.[58]

The contentious politics research program, a subfield of social movement research, primarily focuses on studies of emergent behavior.[59] As a result, much of the research on resistance movements addresses questions surrounding radicalization and mobilization processes, or how and why an organization emerged or gained ground against its rivals and targets. This means that less attention has been devoted to the important issue of movement demobilization in the body of research. One of the difficulties in researching movement decline is a failure to agree on conceptual terminology to describe the end state of a resistance movement. Christian Davenport notes that the literature addressing the issue across different disciplines refers to a movement's declines using widely divergent terms describing the same phenomenon, including *decline, contraction, (dis)continuity, termination, disbanding, decay, demobilization,* and *outcome,* among others.[60] This section discusses numerous widely used and accepted terms to describe different forms of movement decline.

Operational decline in resistance movements looks different depending on numerous factors.[61, 62 63] According to Davenport, there are two primary instances when decline occurs. Decline occurs externally when pressures are exerted on the movement, usually some sort of repression initiated by state authorities. However, demobilization can also occur when pressures from inside of the group, such as fractionalization or ideational rigidity, dismantle the group internally. Sometimes, however, the external pressures leveraged by state authorities against resistance movements function through these internal mechanisms.[64]

Among resistance movements that are party to a civil war, a resolution can signal a defeat by government forces, a victory, or a negotiated settlement. Sometimes, the resolution appears more informal, such as an undeclared cease-fire or stalemate that signals that an organization has simply fizzled out into dormancy or exhaustion. Resistance may or may not be rekindled later in a similar or substantively transformed manner (e.g., different name, different goals, and different leaders). However, among nonviolent, mass-based resistance movements, demobilization can mean that the organization has been co-opted or mainstreamed by its opponents. If a movement is co-opted, its leaders, participants, or both are invited to join the political process, moving from the periphery to the center.[65, 66] In this manner, co-optation decreases the need and incentives for further participation in the movement.

Alternatively, mainstream elites sometimes adopt the goals and rhetoric of the movement to sap support from the movement, a process that Davenport calls "problem depletion."[67, 68] When problem depletion occurs, the incentives for further participation in the movement are considerably diminished, often resulting in the decline of the group.[69] In the dynamic between the resistance movement and its target state, the state's repressive actions can successfully dismantle incentives for participants and deplete the movement's resources, resulting in the death of the movement. Internal dynamics, sometimes receiving a strategic push from state actors, can also result in the implosion of a movement. The following discussion of these various resolutions or declines is organized in descending order from forms of decline that are generally positive for participants in an opposition or resistance movement to those that are negative for members.[70]

SUCCESS

Success is more complicated than other forms of decline.[71] Although one could imagine a resistance that sets particular goals, achieves them, and then subsides, it is more common for movements to be "forced into compromises that only sometimes are advantageous to the movement."[72] Although it obtains concessions from its opponent, the movement generally still relinquishes or compromises on some portion of its original claims.[73] This dynamic of absorption soon transforms what was once a resistance movement instead into an interest group, brought into the conventional political institutions of the state.[74]

The shape of success and the concessions required can reveal internal fractures within a movement. Competing priorities and conceptions of what success means can lead to division, radical offshoots, and loss of legitimacy in the eyes of "true believers," all likewise causing impotency and decline. The radical offshoots can be "spoilers" in the peace process; in the case of violent resistance movements, this can be a continuation of armed attacks. Some members of the resistance movement may see success when certain goals are achieved, but others may perceive success only when the movement continues to grow, expanding its priorities beyond the accomplished short-term goals. However, growth may also lead to the attraction of new members who are less committed to the original resistance than older members. The new generation of members may lead to factions that weaken the movement overall or change the nature of the movement entirely.

Ukraine's Orange Revolution: Decline through Success

The Orange Revolution is an example of a resistance movement that transitioned to the resolution state through success. Resolution through success indicates some degree of fulfillment of resistance goals, but it also indicates the decline of the resistance in response to those successes. A popular uprising against the fraudulent election of pro-Russian presidential candidate Viktor Yanukovich, the Orange Revolution started to transition from escalated protests and conflict toward decline when a third election took place on December 26, 2004. Viktor Yushchenko, the pro-Western candidate, won the new ballot by a clear margin. After a prolonged legal battle waged by Yanukovich, the Supreme Court upheld the new results, and Yushchenko was sworn in as Ukraine's president on January 23, 2005, signaling the successful resolution of the movement. The movement declined and splintered even further after the decisive election because it lacked a unifying enemy, leading to infighting between factions that paved the way for Yanukovich to reemerge and be elected president in 2010, defeating former movement leader Yulia Tymoshenko. Some view the 2014 Euromaidan uprisings in Kiev as the reemergence of this movement, but the successful ouster of Yanukovich, combined with ongoing conflict in eastern regions of the country, likewise saw internal division and movement decline.

Chuck Crossett (ed.), *Casebook on Insurgency and Revolutionary Warfare, Volume II: 1962–2009* (Fort Bragg, NC: USASOC, 2012), 625–643; W. Sam Lauber, Steven Babin, Katherine Burnett, Jonathon Cosgrove, Theodore Plettner, and Catherine Kane, *Understanding States of Resistance*, Draft (Fort Bragg, NC: USA-SOC, 2016), 47, 98–99.

Facilitation

Facilitation occurs when the government or vested interests bring about the decline of a resistance group or movement through the satisfaction of at least some of its claims.[75] Opposite but related to repression, facilitation may be pursued to a limited degree and combined with measures of repression.[76] When the government facilitates some, but not necessarily all, of a resistance group's claims, this may have the effect of splitting the resistance movement. Facilitation may attract movement moderates away from resistance activities and toward legitimate action, frustrating radicals who want more change. Such a split will weaken the resistance as wings satisfied with facilitation will call for de-escalation, while holdouts persist in aggressive opposition despite government compromises, possibly causing a decline in popular support. Facilitation is often used by governments in coordination with selective measures of repression as an effective means to end a resistance movement.

Provisional Irish Republican Army: Decline through Facilitation

The PIRA declined into a state of resolution on April 10, 1998, through a process of facilitation. The Good Friday Agreement in 1998 satisfied some of the PIRA's demands, including policing reforms, the release of political prisoners, provisions for a popular vote on Northern Ireland's status, and establishment of power-sharing institutions. After the agreement, Sinn Féin, the political arm of PIRA, became one of the largest parties in Northern Ireland. The agreement also disarmed the PIRA, and in 2005, international observers announced its complete demobilization. Popular support for the agreement was displayed in votes in favor of the resolution in 1999 (71 percent of voters in Northern Ireland and 94 percent in Ireland). Despite these achievements, the movement's primary goal of an independent and unified Ireland was not met, but the insurgency nevertheless declined to resolution.

Crossett, ed., *Casebook on Insurgency and Revolutionary Warfare*, 293–327; Lauber et al., *Understanding States of Resistance*, 46, 91–92.

Institutionalization

Also referred to as bureaucratization, institutionalization occurs when a resistance movement adopts less-extreme ideologies and replaces disruption with more conventional forms of contention.[77] During the process of institutionalization, the movement's leaders actively seek points of compromise or accommodation with the political regime, generally through moderating its own goals.[78] Hopper referred to this form of decline as "the institutional stage of legalization and societal organization," where the group transforms itself into a permanent organization that is more acceptable than unacceptable to mainstream society.[79] Institutionalization can often simultaneously occur with radicalization among the portion of the movement that rejects the moderate direction of the larger movement. While both institutionalization and radicalization lead to decline in a movement, the former is a partial success of the movement insomuch as when institutionalization occurs the movement becomes a more permanent voice in the status quo, although not fully accepted within it. Depending on the perceived extent of this success, the resistance movement may lose its primary motivating force.

> *Institutionalization occurs when a resistance movement adopts less-extreme ideologies and replaces disruption with more conventional forms of contention.*

Palestine Liberation Organization: Decline through Institutionalization

The PLO offers an example of decline through gradual institutionalization. After the first intifada in 1987, the PLO began moderating its tactics and shifting from armed resistance to diplomacy and governance, most notably through its recognition of Israeli statehood, participation in the 1993 Oslo Accords, and creation of the Palestinian Authority. This moderation of tactics led to a decline in popular support over time, allowing the more radical Hamas to gain footing among the public. This shift in popular support to the more radical Hamas was apparent by its electoral control of the Palestinian Legislative Council after 2006 elections and the violent takeover of the Gaza Strip in 2007. Hamas's victory in the January 2006 elections signaled the PLO's transition into decline as an institutionalized group rather than a subversive and resistance-oriented one. The PLO continued to act as a representative of the Palestine movement, especially among international audiences, but largely through diplomatic, institutionalized channels rather than armed resistance.

Crossett, ed., *Casebook on Insurgency and Revolutionary Warfare*, 213–235; Lauber et al., *Understanding States of Resistance*, 45, 88–89.

Establishment with the Mainstream

Decline through mainstream acceptance is the process through which a movement becomes a fully accepted part of the legitimate political system. Movements that establish with the mainstream generally realize most of their goals, which means they no longer have to challenge the status quo.[80] Although it is similar to institutionalization, establishment with the mainstream denotes gaining acceptance as a voice within the dominant power structure while simultaneously avoiding co-optation by existing powerful interests.

FMLN: Decline through Establishment with the Mainstream

The FMLN in El Salvador provides a useful example of decline through establishment with the mainstream. The FMLN signed a peace accord with the government of El Salvador on January 16, 1992. Peace negotiations leading up to the 1992 accord would have been unlikely without the growing influence of moderates within the FMLN ranks who saw violence as unsustainable and unlikely to bring victory, alongside growing exhaustion among the landed elite who suffered economically during the civil war. By signing the accord, the FMLN accepted concessions from the government, most notably gaining recognition as a political party, allowing the leadership of the group to enter the mainstream. The accord addressed other critical demands of the FMLN, including land reforms to help the peasant class, the creation of an independent body (the UN Truth Commission for El Salvador) to investigate atrocities carried out during the war, and the establishment of a civilian police force and constitutional limits on the military's power. Lastly, the accord outlined the demobilization of both the FMLN and the Armed Forces of El Salvador, which was carried out under UN observation over eighteen months after the signing of the accord. Today, the FMLN operates as one of the largest parties in El Salvador, effectively declining from a movement of resistance and emerging as mainstream.

Crossett, ed., *Casebook on Insurgency and Revolutionary Warfare*, 117–148; Lauber et al., *Understanding States of Resistance*, 48-49, 86–87.

Co-Optation

State authorities battling a resistance movement often adopt a strategy of co-optation to force the movement into decline. Co-optation is a strategy that provides movement leaders with rewards that advance their private interests at the expense of the collective good of the larger movement.[81] The rewards or positions are meant to ensure that resistance leaders see a convergence of their interests with those of the political and economic elite in society.[82] Movements that are especially vulnerable to this strategy are those with highly dependent or centralized leadership and movements formed around charismatic leaders,[83] as co-optation of that leader into a lucrative position effectively disarms the group of a primary motivating factor.

Co-optation is a state-driven strategy that provides resistance movement leaders with rewards that advance their private interests at the expense of the collective good of the larger movement.

Scholars have also described the co-optation of a movement (particularly social movements) in terms of a process. First, state authorities appropriate the means used by the movement, including its narrative and forms of contention.[84] Second, authorities assimilate leadership and members through limited inclusion and participation in the legitimate political system, particularly on decisions regarding the issues that most concern the movement. Third, authorities attempt to influence the movement's goals so that they more closely align with the incumbent government's interests. With greater influence, authorities are well suited to regulate and control the movement to serve the interests of the state.[85]

Abeyance

Although it does not mark the end of a movement, abeyance is a state of decline that is sometimes called dormancy.[86] In this state, the movement is not actively engaged in mobilization but focuses inward on its own identity.[87] This inactive period is also marked by reduced recruitment and the avoidance of confrontations with adversaries. According to sociologists Traci Sawyers and David Meyer:

> During abeyance, movements sustain themselves but are less visible in interaction with authorities. At the same time, values, identity, and political vision can be sustained through internal structures that permit organizations to maintain a small, committed core of activists and focus on internally oriented activities.[88]

Similarly, another sociologist theorizes that movement abeyance provides continuity for the group while allowing time to build support even while confronted with a hostile or unreceptive political authority.[89] Resistance movements, despite falling back into a dormant state through abeyance, can reemerge and remobilize after reinforcing group identity and developing a larger support base.

Abeyance is a period of dormancy or inactivity in a resistance movement. When in abeyance, a resistance movement does not actively mobilize the population but focuses on internally related matters, such as identity or organization.

Radicalization

A movement begins to decline when it radicalizes by shifting toward ideological extremes or adopting escalating disruptive and/or violent forms of contention.[90, 91] Radicalization does not necessarily cause decline, but it can serve as a mechanism for demobilization, particularly when one wing of the movement radicalizes in reaction to the moderation, co-optation, or institutionalization of a rival wing.[92] If one wing of the movement is declining through compromise and capitulation while the other radicalizes even further into nonnegotiable positions and escalatory violence, the movement may lose legitimacy, crippling recruitment and mobilization efforts until the radical wing declines into irrelevance.

Radicalization occurs when members or leaders of a resistance movement shift toward ideological extremes or adopt increasingly disruptive or violent tactics.

Exhaustion

After a resistance movement has matured and weathered a long struggle, the organization and other participants may experience gradual decline as psychological exhaustion damages the emotional momentum of the group.[93] This slow deflation of zeal for resistance through exhaustion occurs through the eventual success of moderate or established interests that advocate for decline and a return to normalcy. Davenport calls this exhaustion "burnout" and notes that activism is hard work physically, emotionally, and psychologically. As a result, many activists "will just get fed up and quit."[94] While burnout or exhaustion is a very real phenomenon, few have systematically investigated this important form of demobilization. Scholars in contentious politics outline exhaustion as a state of decline: "although street protests, demonstrations, and violence are exhilarating at first . . . [resistance movements] involve risk, personal costs, and, eventually, weariness and disillusionment."[95] Exhaustion can cause unequal decline in participation and contribute to internal moves toward radicalization or institutionalization, which are efforts to respectively either reenergize or salvage the influence of a movement in decline.[96]

Repression

The government may succeed in repressing the resistance movement, a state of decline where those in power "use force to prevent movement organizations from functioning or prevent people from joining the movement organizations."[97] Tactics for repression are numerous, including imprisonment, execution, torture, mass killings, infiltration, harassment, threats to job and school access, the spread of false information, and "anything else that makes it more difficult for the movement to put its views before relevant audiences."[98] There are differing reports regarding the impact of repression on the viability of resistance movements. Some research indicates that repression can increase mobilization, while other research finds evidence for the opposite, that repression constrains or neutralizes mobilization.[99, 100] The relationship between repression and political violence is referred to as the conflict-repression nexus.[101]

When repression is successful, it can create an internal and external push toward demobilization. Repression can lead to an internal push for demobilization as repression generates distrust and paranoia among the targeted resistance group, hindering collective cooperation. The effects are likely to be more pronounced according to certain organizational characteristics of the resistance movement. Organizations with highly selective recruitment strategies, for instance, are more likely to experience a higher level of trust that is not as affected by repressive activities that seek to break down the internal cohesion of the movement, particularly those related to wiretaps, informants, and agent provocateurs, or at least suspicion of those measures within the targeted movement. Moreover, resistance movements with experienced activists are likely to have more accurately assessed the government's expected behavior, such as repressive countermeasures, and identified best practices to subvert the effects of such measures on the organization. However, when resistance actors encounter unexpected repression for which they have not prepared, maintaining levels of trust and momentum among the group is more difficult, making the group more prone to demobilization. When resistance organizations are unable to rapidly and effectively adapt to unexpected conditions, demobilization is again more likely. While repressive actions could lead to the decline and resolution of a resistance movement, they can also push radicals into smaller, more exclusive groups and can contribute to the adoption of escalatory violence.[102]

FAILURE

Resistance movements can also decline and fail to achieve their goals for internal reasons, especially in the organization leading the movement. In such cases, the choices or activities of the group itself, rather than overpowering external factors, are the cause of the decline. The internal collapse and failure of the leading movement organization can lead either to movement abeyance or to final decline if no other groups take up the torch. There are two forms of failures: factionalism and encapsulation. In factionalism, the inability of the organization's members to agree on the best direction to take leads to an internal division that ultimately kills the group's momentum. Second, encapsulation occurs when a movement develops an ideology that interferes with its ability to recruit, eventually causing a critical decline in mobilization and capabilities.[103]

Movements can also fail in their infancy and incipience. Scholars have noted four ways in which a movement may mishandle its nascent development, leading to burnout. First, groups can fail by not establishing communication links between the movement and the segment of the population most sympathetic to its goals, which effectively hamstrings the group's growth into a more mature movement.[104] Second, a young group might fail to connect with community leaders and stakeholders, another death sentence for a young movement.[105] Third, the group might lack a sufficiently broad platform so that a potentially sympathetic audience cannot grant the movement its wholehearted support, stifling recruitment and growth.[106] Finally, failures that become highly publicized and conspicuous can fatally weaken the public image of the movement, discredit causing it to rapidly decline or experience a slow, inexorable decline as confidence in the group does not recover.[107]

The Republic of New Africa

The Republic of New Africa (RNA) was a black secessionist movement formed in the United States in the late 1960s. It first emerged at a black nationalist conference held by the Malcom X Society in 1968. Conference participants formed the RNA, basing their platform on demands for a separate black homeland in the United States and reparations for centuries of slavery that had denied black slaves financial compensation for their labor. The group had overt socialist leanings, arguing that the state should hold the major means of production in trust for the people. The RNA issued a declaration of independence and established a governing body for the new black nation. In effect, the RNA formed a new government and then went about the business of getting support for it among the black community through press conferences, rallies, demonstrations, and other disruptive events to garner media coverage. While the group decided on a nonviolent strategy at the outset, it did not rule out the use of violence if it felt threatened or its progress was stymied. The bulk of RNA activists lived in and around Detroit.

Although the RNA worried about repression by police and the Federal Bureau of Investigation, security forces engaged in little repressive behavior at the outset. Peopled with experienced activists, the RNA trained some members in a quasi-armed wing, claiming to have aboveground and underground forces ready to deploy guerrilla tactics in US cities. The formation of the armed groups was in preparation for violent repression, particularly by police, that had plagued black activism during the civil rights movement. The development of the armed wing, a precautionary measure designed to cultivate trust among RNA supporters who were fearful for their lives, was the catalyst that drew more attention—and repression—from the security forces.

In addition, the leadership of the movement argued over its first campaign, the attempted secession of a neighborhood in Brooklyn. The argument created divisions within the RNA, with some arguing for a southern, not a northern, orientation, focusing on the states in the Deep South, notably Mississippi. The divisions were exacerbated after police staged a mass raid against the group, imprisoning many key members and overwhelming its capabilities. Eventually, the existing internal problems within the movement, combined with the weakened trust among RNA members after the police covertly infiltrated the group, led to the development of two factions within the RNA. One faction argued that the group should step away from radical activity that invited repression, while the competing faction argued for stepping up disruptive and radical activity, including the use of violence. Further factionalization resulted from the imprisonment of most of the RNA's leaders in Mississippi. The disagreement and distrust among the organization contributed to the failure of the movement to achieve any of its goals or move substantially past the incipient or crisis phase of resistance.

Christian Davenport, *How Social Movements Die: Repression and Demobilization of the Republic of New Africa* (*New York: Cambridge University Press, 2015*).

CONCLUSION

This chapter discussed conflict termination and demobilization, including its implications for a mission's strategic outcomes. Most often, the objectives of military missions focus on near-term outcomes. In the case of unconventional warfare, these objectives might include the overthrow or disruption of an incumbent regime. However, in the interest of securing enduring goals, it is necessary to have a clearer understanding of the dynamics in a state as a resistance movement declines. The postconflict environment is replete with challenges that threaten long-term strategic goals. States that experience civil war have difficulty consolidating democracies and are at high risk of experiencing another civil war. Political scientists were probably among the least surprised that the fragile peace in Iraq disintegrated after the withdrawal of US forces. Civil wars that end in negotiated settlements, which third parties often design as good political solutions for themselves, are especially vulnerable to recidivism.

The types of termination or demobilization are stratified into studies on civil wars and social movements. In civil wars, wars terminate via negotiated settlements, military victories, cease-fires, or informal declines in violence. In the post-Cold War era, more civil wars ended through negotiated settlements, usually brokered by third parties, than military victories. This poses some difficult challenges in the postconflict environment because negotiated settlements are the termination type most associated with a recurrence of civil war and violence.

In contrast to the research on violent resistance movements in studies of civil war, social movement theory provides a rich conceptual spectrum of demobilization. Resistance movements can decline through mostly positive processes, such as facilitation or institutionalization, or negative processes, such as co-optation and repression, that inhibit the capacity of the group to continue operations. As a resistance movement interacts with state security forces, factions within the movement or within competing resistance movements make choices that impact the maturation and eventual decline of the movement.

ENDNOTES

[1] William Flavin, "Planning for Conflict Termination and Post-Conflict Success," *Parameters* 33, no. 3 (Autumn 2003): 96.

[2] Quantitative research studies using statistical regression methods are often called "large-n" studies because the validity of the results is based partially on the large number of cases included in the analysis.

[3] Peter Wallenstein and Margareta Sollenberg, "Armed Conflicts, Conflict Termination and Peace Agreements, 1989–96," *Journal of Peace Research* 34, no. 3 (1997): 342.

[4] Gordon H. McCormick, Steven B. Horton, and Lauren A. Harrison, "Things Fall Apart: The Endgame Dynamics of Internal Wars," *Third World Quarterly* 28, no. 2 (2007): 324.

[5] In the database, the researchers note that there are several loose forms of cease-fire agreements. In some cases, cease-fires are agreements short of negotiated settlements, with the parties to the conflict reaching an informal agreement to maintain the status quo with little chance that fighting will resume. Other cease-fires are short breaks in the fighting, and observers expect fighting to resume. Sometimes, levels of fighting among the parties de-escalate to low thresholds that do not meet the twenty-five battle death threshold. In some of these cases, the de-escalation signals a near victory for the dominant side.

[6] Wallenstein and Sollenberg, "Armed Conflicts," 342.

[7] Joakim Kreutz, "How and When Armed Conflicts End: Introducing the UCDP Conflict Termination Dataset," *Journal of Peace Research* 47, no. 2 (2010): 243.

8 Robert D. Kaplan, *The Coming Anarchy: Shattering the Dreams of the Post Cold War*, (New York: Random House, 2002).

9 Wallensteen and Sollenberg, "Armed Conflicts."

10 Internationalized civil wars are similar to proxy wars.

11 Therése Pettersson and Peter Wallensteen, "Armed Conflicts," *Journal of Peace Research* 52, no. 4 (2015): 536–550.

12 Monica Duffy Toft, *Securing the Peace: The Durable Settlement of Civil Wars* (Princeton: Princeton University Press, 2010).

13 Kreutz, "How and When Armed Conflicts End," 243–250.

14 Toft, *Securing the Peace*, 64–65.

15 Erica Chenoweth and Maria J. Stephan, *Why Civil Resistance Works: The Strategic Logic of Nonviolent Conflict* (New York: Columbia University Press, 201), 201–219.

16 James D. Fearon and David D. Laitin, "Civil War Termination," draft, Paper presented at the Annual Meetings of the American Political Science Association, Chicago, IL, 30 August–1 September 2007: 3, 16.

17 McCormick, Horton, and Harrison, "Things Fall Apart," 346.

18 Stephen E. Gent, "Going in When It Counts: Military Intervention and the Outcome of Civil Conflicts," *International Studies Quarterly* 52, no. 4 (2008): 731.

19 Karl R. DeRouen, Jr. and David Sobek, "The Dynamics of Civil War Duration and Outcome," *Journal of Peace Research* 41, no. 3 (2004): 317–318.

20 David E. Cunningham, "Blocking Resolution: How External States can Prolong Civil Wars," *Journal of Peace Research* 47, no. 2 (2010): 115–127.

21 Gent, "Going in When It Counts," 730–731.

22 Fearon and Laitin, "Civil War Termination," 16–17.

23 McCormick, Horton, and Harrison, "Things Fall Apart," 327, 346.

24 DeRouen and Sobek, "The Dynamics of Civil War," 317–318.

25 Gent, "Going in When It Counts," 730–731.

26 Toft, *Securing the Peace*, 151–152.

27 Fearon and Laitin, "Civil War Termination," 1–3.

28 DeRouen and Sobek, "The Dynamics of Civil War," 317.

29 J. Michael Quinn, T. David Mason, and Mehmet Gurses, "Sustaining the Peace: Determinants of Civil War Recurrence," *International Interactions* 33, no. 2 (2007): 188.

30 T. David Mason, Mehmet Gurses, Patrick T. Brandt, and Jason Michael Quinn, "When Civil Wars Recur: Conditions for Durable Peace after Civil Wars," *International Studies Perspectives* 12, no. 2 (2011), 186–187.

31 Ibid.

32 Caroline Hartzell and Matthew Hoddie, "Institutionalizing Peace: Power Sharing and Post-Civil War Conflict Management," *American Journal of Political Science* 47, no. 2 (2003): 323, 328.

33 Kreutz, "How and When Armed Conflicts End," 243.

34 Quinn, Mason, and Gurses, "Sustaining the Peace," *International Interactions* 189; Mason, Gurses, Brandt, and Quinn, "When Civil Wars Recur," 186–187; Joakim Kreutz, "How and When Armed Conflicts End: Introducing the UCDP Conflict Data Determination Dataset," *Journal of Peace Research* 47, no. 2 (August 2010): 243.

59 Ruud Koopmans, "Protest in Time and Space: The Evolution of Waves of Contention," in *The Blackwell Companion to Social Movements*, eds. David A. Snow, Sarah A. Soule, and Hanspeter Kriesi (Malden, MA: Blackwell Publishing, 2004), 26.

60 Christian Davenport, *How Social Movements Die: Repression and Demobilization of the Republic of New Africa* (New York: Cambridge University Press, 2015), 21.

61 Davenport identifies four key characteristics of demobilization, several or all of which may be present in empirical cases: "(1) Official termination and/or significant alteration of the formal institution engaged in challenging authorities; (2) departure of individuals (members) from relevant organizations—especially the founding and/or core members that participate most frequently; (3) termination or of significant reduction in dissident interventions (behaviors); and (4) a fundamental shift in the ideas of the challenger (particularities of the claim) away from what was earlier established."

62 Ibid., 21–22.

63 Ibid., 21.

64 Ibid., 21–60.

65 Some scholars do not include these instances of "positive demobilization" in demobilization processes that contribute to the demise of the movement without it having fulfilled significant goals, such as integration into mainstream politics.

66 Ibid., 22.

67 Davenport argues that one of the best examples of problem depletion strategies is in prototypical counterinsurgency practices that seek to win the hearts and minds of the population targeted by the resistance movement.

68 Ibid., 26–28.

69 Mainstreaming, however, can also be viewed as a type of success because some of the resistance movement's demands are met by the political system.

70 Ibid., 38–57.

71 Miller, "The End of SDS and the Emergence of Weatherman," in Jo Freeman, (ed.), *Social Movements of the Sixties and Seventies* (New York: Longman, 1973), 306–307.

72 Ibid.

73 Ibid.

74 Ibid.

75 Facilitation is also sometimes referred to as accommodation.

76 Sidney Tarrow, *Power in Movement: Social Movements and Contentious Politics.* (Cambridge: Cambridge University Press, 1998), 54, 127, 189-190.

77 Ibid., 207-208.

78 Ibid.

79 Rex D. Hopper, "The Revolutionary Process: A Frame of Reference for the Study of Revolutionary Movements," *Social Forces* 28, no. 3 (March, 1950): 270–279.

80 John J. Macionis, *Sociology*, 9th ed. (Upper Saddle River, New Jersey: Prentice Hall, 2003), 619.

81 Miller, "The End of SDS and the Emergence of Weatherman," 305.

82 Ibid.

83 Jonathan Christiansen, "Social Movements & Collective Behavior: Four Stages of Social Movements," *Research Starters Academic Topic Overviews*, EBSCO, 2009, 4.

84 See chapter 6 for a discussion of repertoires of contention.

85 Patrick G. Coy and Timothy Hedeen, "A Stage Model of Social Movement Co-Optation: Community Mediation in The United States," *The Sociological Quarterly* 46, no. 3 (2005): 411, 413–426.

86 Stacy Keogh, "The Survival of Religious Peace Movements: When Mobilization Increases as Political Opportunity Decreases," *Social Compass* 60, no. 4 (2013): 561–578.

87 Christiansen, "Social Movements & Collective Behavior," 6.

88 Traci M. Sawyers and David S. Meyers, "Missed Opportunities: Social Movement Abeyance and Public Policy," *Social Problems* 46, no. 2 (1999): 188.

89 Verta Taylor, "Social Movement Continuity: The Women's Movement in Abeyance," *American Sociological Review* 54, no. 5 (1989): 762.

90 See chapter 6 for a discussion of processes relating to radicalization and pertinent examples.

91 Tarrow, *Power in Movement*, 207.

92 Ibid., 190, 207–208.

93 Hopper, "The Revolutionary Process," 277.

94 Davenport, *How Social Movements Die*, 32–33.

95 Tarrow, *Power in Movement*, 206.

96 Ibid.

97 Miller, "The End of SDS and the Emergence of Weatherman," 304–305.

98 Ibid.

99 Some researchers have found that repression increases the likelihood of resistance mobilization, while others find that repression decreases the incidence of resistance mobilization by closing off opportunities to organize, recruit, and gather resources.

100 See Christian Davenport, "State Repression and Political Order," *Annual Review Political Science* 10 (2007): 1–23; David Ortiz, "Confronting Oppression with Violence: Inequality, Military Infrastructure and Dissident Repression," *Mobilization* 12, no. 3 (2007): 219–238.

101 Mark Irving Lichbach, "Deterrence or Escalation? The Puzzle of Aggregate Studies of Repression and Dissent," *Journal of Conflict Resolution* 31, no. 2 (1987): 266–297.

102 Tarrow, *Power in Movement*, 209.

103 Miller, "The End of SDS and the Emergence of Weatherman," 307–308.

104 Maurice Jackson, Eleanora Petersen, James Bull, Sverre Monsen and Patricia Richmond, "The Failure of an Incipient Social Movement," *The Pacific Sociological Review* 3, no. 1 (Spring, 1960), 40.

105 Ibid.

106 Ibid.

107 Ibid.

CHAPTER 6.
CONCLUSION

The end of the Cold War ushered in an era of irregular conflict that continue to shape future threats in the international security environment. Since the seventeenth century's Peace of Westphalia, sovereign states have dominated international affairs. The past several decades have seen this dominance erode with the emergence of powerful non-state actors on the world stage, as well as the degeneration of sovereignty in so-called failing or weak states. Internal conflicts, or civil wars, have replaced the threat of interstate wars. Across the globe, violent and nonviolent resistance movements are challenging the authority of state governments.

The decrease in interstate wars and the subsequent rise of belligerent nonstate actors has meant that the future threats facing the United States and its allies are no longer limited to conventional armies of near-peer competitors. During the Cold War, conventional warfare remained the predominant paradigm as the United States grappled with its superpower rival, the Soviet Union. Even during that time, however, the United States continued combatting irregular force adversaries. Irregular conflict, characterized by weaker opponents, combatants diffused in local populations, and innovative tactics, is an inextricable component of the threat landscape today and into the future. The US military's emphasis on traditional warfare has given way to the acknowledgement that irregular warfare is no longer an aberration but an essential component of contemporary conflict, resulting in efforts to integrate the precepts of irregular conflict into doctrine and concepts. Indeed, careful study of US military history reveals that the US military has long been involved in irregular warfare, but that observation is often overshadowed by the large, conventional wars of the twentieth century. Combating the complexity inherent within this type of warfare requires cultivating a disciplined, rigorous approach to the science of resistance.

Resistance movements generally attract attention from the military as their leaders transition the movements to violence in their efforts to challenge state authority. However, in the past several decades, nonviolent resistance movements have proven tremendously resourceful and effective, toppling governments or forcing policy reform on incumbent regimes. Movements such as the Polish labor union, Solidarity, or protestors in the Arab Spring accomplished goals most observers thought impossible or highly unlikely—unseating entrenched authoritarian regimes with highly capable security forces. Indeed, research on resistance movements demonstrated that nonviolent campaigns have proven more effective in securing their political goals than their violent counterparts. This observation, combined with the insights previously discussed, should prompt the Special Forces, other military leaders, and policy makers to revisit their concepts of resistance to incorporate these academic developments.

The definition of resistance has evolved over time. Early definitions of resistance, developed largely during Second World War, focused on resistance as a partisan domestic effort against an occupying force, distinguishing it from other types of warfare. Eventually, the concept broadened to include efforts outside of those against a foreign power and encompassed actors that did not rely on the use of violence to achieve their objectives. While the expanded concept is a welcome addition, it is also necessary to ensure that the Special Forces focuses on types of resistance that are operationally relevant. The definition of resistance supplied by the ARIS team seeks to strike the balance between treating the phenomenon holistically while maintaining operational relevance. The definition emphasizes the fundamental asymmetry of resistance, its organizational characteristics, and its intent to subvert and disrupt political targets. As a result, the ARIS definition restricts consideration to movements planning to undermine, thwart, or oust incumbent regimes outside the bounds of "normal" political activity.

The ARIS definition of resistance also points to a broadened conception of resistance that dismisses violence as a discerning feature of the phenomenon. Many violent resistance movements did not use violence at the outset but turned to weapons only after a period of nonviolence. Resistance therefore does not begin when the groups pick up guns or the body count on the battlefield reaches a prescribed threshold. Rather, resistance is a continuous concept in which groups adopt various strategies and tactics over their life cycles, including violence and nonviolence. It is incumbent on the Special Forces, with the aid of the work presented here and in the larger ARIS corpus, to help senior decision makers discern which movements are likely to be a strategic threat or asset.

In addition to providing a semantic definition, the ARIS project also describes a set of attributes common to most resistance movements. The key attributes capture the interactions between a resistance movement and its surrounding environment for a holistic analysis. The attributes include actors, causes, organizations, actions, and environment. The actors attribute nods to the persons or groups indirectly or directly involved in a movement, whether as participants, leaders, or passive and active supporters in the local population. Opportunity captures the environmental conditions that alter incentives for mobilization by minimizing risk or increasing the rewards of actors' participation. The opportunities might be a shift in governmental policies or a demonstration of a regime's vulnerabilities and weaknesses. Operationally relevant resistance occurs within an organization, serving as the platform for collective action. In turn, organization is enabled by other factors, such as the formation of a group identity, a shared ideology, or shared interests. Finally, resistance movements make their presence felt by their opponents through action. Action occurs as the pursuit of common organizational goals encompasses a broad set of strategies and tactics, including both violence and nonviolence.

The work presented here is organized according to "big" thematic questions or puzzles that have inspired decades of social science researchers. Many of the questions are similar to those asked within the Special Forces engaging in Unonventional Warfare or Foreign Internal Defense, although with the intent to explain rather than to distill best practices. Although the researchers adopted different frameworks and perspectives, the key attributes of resistance previously described appear in the analysis of resistance throughout their work. The question or puzzle that heads each chapter has by itself produced a large body of research. As is frequently the case with social science, there is rarely an indisputable consensus on the answers to these complex questions. Instead, the ARIS team aimed to provide an intermediate introduction to the major findings and theories explaining different facets of resistance.

The majority of the research presented here is derived from the social sciences. Most draws from political science and sociology, where the bulk of research on resistance is found. In general, research on resistance in the social sciences is divided into camps. Some study violent conflict, usually civil wars, while others look at nonviolent resistance, such as in studies on social movements. However, more researchers have begun to study resistance as a singular phenomenon, although the research program is still relatively young. Contentious politics, for instance, unites disparate studies of nonviolent and violent resistance into a single field of inquiry. The research draws on the rich, case-study-based material found in social movement studies. It applies the concepts, processes, and mechanisms found in social movement theory to studies of violent resistance movements or movements that have transitioned from nonviolence to violence.

The first thematic puzzle raised in this work related to why some countries experience resistance while others do not. The puzzle is most often treated in research on civil war onset, which relies on quantitative analysis of large datasets, but also from studies of social movements. The former literature relies on macrostructural variables to explain why civil war breaks out in some countries or regions. Low levels of socio-economic development, political marginalization of ethnic groups, and bordering neighbors experiencing warfare all increase the likelihood that a country will experience internal armed conflict.

While the evidence is useful for identifying countries at risk, some of the structural variables, such as rough terrain, are difficult or impossible to alter. Therefore, the ability to translate the evidence to actionable strategies or tactics is limited. In the latter literature, the contentious politics research program explains how political opportunities can alter the incentives of individuals and participants to take part in resistance by minimizing the risks or increasing the expected rewards of participation. While research on macrostructural variables tends to relegate actors into the background, research on political opportunity views political processes as emerging from the interaction between actors and their environment. Political opportunities analyze the impact of regime openness and capacity on group decisions to engage in collective action. The configuration of actors is equally important, particularly when considering nonviolent resistance movements. Leaders of these groups must pay careful attention to the interests of the military to assess the likelihood of widespread military defection or armed repression.

The second thematic question raised relates to one of the most enduring puzzles in resistance studies. While most individuals and groups stay on the sidelines during a conflict, others mobilize to engage in collective action. What separates the two populations, the participants and nonparticipants? Grievances abound, but resistance movements are rare. The puzzle is a problematic one because mobilization is among the most significant challenges facing resistance leaders. It requires leaders to cultivate sophisticated social solutions capable of uniting individuals with different ideas, interests, and goals into an organized whole. The theory of collective action first articulated this challenge of mobilization. Scholars, using selective incentive theory, explained how the barrier to mobilization can be overcome by offering selective incentives to participants, whether in terms of loot or prestige. Other research assesses how organizational factors that contribute to mobilization, including the role of social networks and affiliative factors that encourage individuals to reduce uncertainty through group collective identity. The explanation focuses on questions of identity, downplaying the role of grievances and ideology in mobilization processes. Indeed, some research demonstrates that ideological parity often occurs only after the individual has joined the resistance movement. This research also highlights how mobilization can occur because of "in-process" benefits, such as feeling pleasure in agency, rather than expected payouts at some time in the distant future.

Mobilization theories also shed light on how the process differs in violent and nonviolent resistance movements. The control-collaboration model looks at how interactions between civilians and armed actors in conditions of violence shape mobilization in unexpected ways. When territorial control is contested between two armed actors, the cost of nonparticipation for civilians skyrockets. Regardless of whether an individual actually supports the goals of an insurgent group, collaboration occurs to gain armed protection. In effect, the violence forces people to hide their preferences so it is difficult or impossible to ascertain levels of ideational support for the insurgency. Here, violence is the overriding factor predicting when and where collaboration occurs. The model suggests ways that mobilization differs for violent and nonviolent movements. Nonviolent movements have a mobilization advantage that most violent ones do not. Compared with the dynamics generated under conditions of violence, participation in nonviolent movements is less costly and disruptive. Most people are willing to sign ballots and participate in demonstrations. The mass mobilization that occurs in nonviolent movements also invites participation across numerous sections of society, potentially creating links with the political and military elite while garnering more international support.

In military parlance, battles are either won or lost. However, research on conflict termination or movement demobilization reveals a more nuanced landscape than suggested by the win–lose dichotomy. Research on civil wars demonstrates that while some of these conflicts are still settled by rebel or state military victories, a near parity of conflicts are terminated through negotiated settlements. The latter settlements increased in frequency during the post-Cold War era as more third parties, such as the United Nations, are helping armed actors resolve conflicts through bargaining and aiding in the enforcement of the resulting agreements. Research also demonstrates that the way a conflict ends impacts the chances that the civil war will reignite. Negotiated settlements are particularly prone to recidivism because the settlements generate considerable commitment problems, lack of trust, and uncertainty. Because one of the missions of the Special Forces is to support indigenous resistance movements through UW, it is especially important to gain a more comprehensive understanding of the attributes and qualities of resistance movements that contribute to a more lasting and stable peace.

Research from social movement theory provides a more nuanced look at how nonviolent resistance movements end. These movements may not simply succeed or fail but instead experience a spectrum of positive and negative outcomes. Movements might be co-opted by the incumbent regimes or establish themselves within the legitimate, mainstream political system. Others might simply fade out due to exhaustion or after experiencing repression by state security forces. While the failure of a resistance movement is generally perceived as resulting from interactions with its opponents, a movement's demise might also be initiated by internal dynamics. Movements are rarely, if ever, monolithic. Ideological, strategic, and personal differences among the leadership and rank and file can result in an internal implosion, particularly as the movement attempts to recover and respond to external repressive measures from the state, competing movements, and counter movements.

One of the pressing questions motivating the ARIS project regards the decision of resistance leaders to transition from nonviolence to violence. Some resistance movements rely on predominantly nonviolent tactics, while others rely on violence to press their political claims. Most often, the framework for explaining the transition relies on rationalist calculation where the movement adopts violence as the most effective strategy available at that time. However, as research on nonviolent movements demonstrates, violence can be a suboptimal outcome. Research from the contentious politics program, by contrast, looks at violence as an emergent phenomenon arising from repeated interactions between opponents. Most often, mechanisms involving competition are the spur toward violence.

Other approaches to the question, however, reveal how violence emerges from other dynamics, including interactions with other actors inside and outside the movement. The interactions prompt mechanisms related to competition over scarce resources that can propel a movement to use violence. This confounds the view that the adoption of violence is a sudden transition, but more likely resides in a gradual escalation. Competition occurs between the resistance actor and the incumbent regime. Resistance movements, particularly nonviolent ones, must maintain heightened levels of disruption that require continual tactical innovation. Security forces respond in kind, generating tit-for-tat dynamics that contribute to a gradual transition to more violent forms of collective action.

While most often observers note the competition with the incumbent regime, resistance actors also face stiff competition internally and from similar groups. Resistance actors are not monolithic but are comprised of a dense web of competing interests, goals, and ideas with more moderate and radical contingents. The differences become especially salient when contestation emerges over how to respond to external pressure from security forces. The more radical contingent is liable to split off from the moderate core, leading to the formation of a violent splinter group. Competition also emerges when groups in the same, usually organizationally dense, movement compete with one another over scarce resources available from popular or international support. When one group adopts a radicalized stance, other groups must adopt similar stances to maintain their legitimacy as a representative of their constituencies. Referred to as political outbidding, the competition pulls the groups toward more radical or violent tactics.

The interactive landscape is not the only factor impacting a movement's strategy and tactics. Certain organizational structures are also prone to more or less violent behavior. Rigid, hierarchical groups, evidencing strong command and control, are capable of wielding extended and intense violence against their opponents. A group's resource endowments, another structural factor, also influence the type of violence a group is likely to wield. Predatory groups that attract participants through financial gain rely more on indiscriminate violence against civilian populations. Similarly, the extent of a movement's territorial control affects the type and lethality of its violence. Clandestine groups, with limited or no territorial control, are far more likely to rely on terrorist tactics. In full-blown insurgencies with areas of contested control, the disadvantaged armed actor relies more on indiscriminate violence due to information and identification problems inherent within asymmetric warfare. As an actor gains control of territory, its levels of violence taper down as information and identification difficulties dwindle.

Since its inception, the ARIS project has sought to further expand the science of resistance, treating it as a singular phenomenon observable in various settings, times, and manifestations. The ARIS research team has continually sought to push the boundaries of resistance scholarship while remaining operationally relevant to its core audience. Students of resistance, whether as observers or practitioners, benefit from exposure to the rich literature already available on the subject within social science scholarship. This work provides a grounded introduction to the material, although by no means exhaustive, enabling more rigorous reflection and analysis of additional existing scholarship on this critical subject.

APPENDIX

THEORIES RELATED TO THE SCIENCE OF RESISTANCE

Abeyance

Abeyance is a period of dormancy or inactivity in a resistance movement. When in abeyance, a resistance movement does not actively mobilize the population but focuses on internally related matters, such as identity or organization. Although recruitment is reduced, a small core of activists sustains the integrity of the organization. During a period of abeyance, a resistance movement also avoids confrontations with its adversaries. A resistance movement can return to a state of active mobilization following abeyance.[1]

Affiliative Factors and Social Networks

Affiliative factors describe how the emotional needs for belonging and social interaction can facilitate mobilization into resistance movements. When the needs remain unfulfilled, individuals may be more susceptible to joining a resistance movement after integrating into a social network for affiliative fulfillment that includes radicalized members.[2]

Anocracies, Hybrid Regimes, and Illiberal Democracies

Anocracies, hybrid regimes, or illiberal democracies share features common to both authoritarian and democratic governments. For instance, an anocratic regime may allow opposition political parties to form and participate in elections but rig elections so that the ruling party is never seriously challenged. Anocracies are also described as states with weak central governments lacking effective policing and counterinsurgent components. Measures of anocracy are frequently referenced from the Polity Project dataset, which measures the qualities of democratic and authoritarian regimes for each country in the world. Some research indicates that anocracies are more prone to political violence than either democratic or authoritarian regimes, the so-called "U-shaped" or curvilinear relationships between regime type and political violence, although other research has challenged these findings.[3]

Bad Neighborhoods

Countries that share borders with states experiencing a civil war are more likely to experience a civil war themselves. These regions are called "bad neighborhoods." The effect is likely due to flows of refugees but also psychological processes that lower natural inhibitions toward violence.[4]

Conflict Trap

The conflict trap refers to the tendency for countries that have experienced one civil war to break down into violent conflict after a period of peace. Several factors contribute to the conflict trap, including the ready availability of weapons and fighters but also the poor economic and social conditions after a country's infrastructure and people are ravaged by war.[5]

Control-Collaboration Model

The control-collaboration model incorporates the interactions between civilians and armed actors to enable a better understanding of mobilization processes. The model applies the same basic logic of the collective action framework but focuses on how the dynamics of violence, irregular warfare, and territorial control impact the mobilization preferences of individuals in affected communities. In the model, mobilization is more likely in areas of contested territorial control under conditions of irregular warfare because individuals have strong incentives to seek out the protection offered by an armed group.[6]

Co-optation

Co-optation is a state-driven strategy that provides resistance movement leaders with rewards that advance their private interests at the expense of the collective good of the larger movement. The rewards or positions are meant to ensure that resistance leaders see a convergence of their interests with those of the political and economic elite in society. In the co-optation process, states appropriate the movement's narrative, assimilate its leaders and members, and thereby attempt to influence the movement's goals so that they align with the interests of the incumbent regime.[7]

Different Pace of Demobilization

A different pace of demobilization occurs when state repression and negotiation raises the cost of participation for members or resolves pressing issues. As a result, moderates and peripheral participants leave the organization, leaving a core of highly committed, radicalized members that increases the likelihood the resistance movement will transition to violence.[8]

Economic Opportunity

Economic opportunity is used to explain mobilization into resistance groups. When economic development in an area is poor, people are more likely to join a movement because it provides the best economic payoff available. This would not be the case in areas of higher economic development because the benefits of participation in a resistance movement are likely to be lower than economic payoffs in the formal economy. Highly skilled technical workers in Silicon Valley, for instance, are not likely to economically benefit more from participation in a resistance movement because employment in the area is so lucrative.[9]

Establishment with the Mainstream

Establishment with the mainstream is a decline process through which a resistance movement is incorporated into the mainstream or legitimate political system. Movements that establish with the mainstream generally realize most of their goals, which means they no longer have to challenge the status quo.[10]

Ethnic Minority Rule

Ethnic minority rule occurs when a minority ethnic group rules over a majority ethnic group(s) through means of political exclusion.

The EPR dataset measures the ethnic minority rule and political exclusion.[11]

Ethnic Polarization Model

The ethnic polarization model argues that ethnic violence does not result from an increased number of ethnic groups in society; diversity itself does not drive political violence among ethnic groups. Instead, it matters how the ethnic groups are configured in society. When a society has a large, majority ethnic group, with smaller, peripheral minority groups, violence between the groups is more probable.[12]

Facilitation

Facilitation is a demobilization process in which the incumbent government or other vested actors bring about the decline of a resistance group or movement through the satisfaction of at least some of its claims. When the government facilitates some, but not necessarily all, of a resistance group's claims, this may have the effect of splitting the resistance movement. Facilitation may attract movement moderates away from resistance activities and toward legitimate action, frustrating radicals who want more change. Such a split will weaken the resistance as wings satisfied with facilitation will call for de-escalation, while hold-outs persist in aggressive opposition despite government compromises, possibly causing a decline in popular support. Facilitation is often used by governments in coordination with selective measures of repression as an effective means to end a resistance movement.[13]

Horizontal Inequality

Horizontal inequality argues that ethnic groups that experience systemic political and economic exclusion as a group (in comparison with other groups in society) are more likely to engage in armed rebellion than others.[14] Horizontal inequality is similar to relative deprivation, but the former looks at group-level inequality while the latter looks at individual levels of inequality. Research conducted using this theoretical model combines measurements to capture inequality and ethnic settlement patterns by using geocoded inequality and ethnic settlement data.[15, 16]

Ideology

An ideology is a comprehensive set of interrelated beliefs, values, and norms. Every society shares commonly held cultural beliefs, including ideas, knowledge, lore, superstitions, myths, and legends. The beliefs in turn are associated with values or judgements of right or wrong that guide individual action. This code is reinforced through a system of rewards and punishments so that approved patterns of behavior, or norms, can discipline the behavior of the group. Individuals seek to give meaning and organization to unexplained events through generalized beliefs like ideology. Movement leaders can interpret situations in terms of the group's beliefs or ideology, translating abstract, ideological beliefs into specific, concrete collective action.[17]

Injustice Frames

Injustice frames are interpretations proffered by movement leaders that highlight how adversaries are actively bringing about suffering or harm to affected groups. When successful, injustice frames help to ignite emotional responses, including hot cognition, that facilitates participation or support with the group.

In-Process Benefits

In-process benefits are the emotional benefits a person experiences while participating in a resistance movement that can serve as a motivation for joining. In this case, participants take pride and pleasure in standing up for their rights and re-asserting their dignity.[18]

Institutionalization

Institutionalization occurs when a resistance movement adopts less extreme ideologies and replaces disruption with more conventional forms of contention. During the process of institutionalization, the movement's leaders actively seek points of compromise or accommodation with the political regime, generally through moderating its own goals.[19]

Legitimacy and Social-Eudaemonic Legitimacy

Legitimacy is the generalized and normative support for an incumbent state authority among the relevant population. Legitimacy may stem from different sources, including tradition, charismatic leadership, and legal-rational procedures such as elections. Another basis of authority is called social-eudaemonic, which means that legitimacy stems from the performance of a government in meeting the demands of its population in terms of the provision of public services. The services may include security, health care, education, sanitation, justice, and economic development.[20]

Loss of Strength Gradient

The LSG predicts that states have less capacity to assert their power the farther away from the state capital. As a result, the state has presence in peripheral regions, providing an opportunity for rebels to emerge and grow. The LSG captures the importance of poor military presence in peripheral regions but also a lack of state capacity to provide adequate social provisions, such as education and health care, that might also contribute to the emergence of violence.[21]

Micromobilization Processes

Micromobilization processes are discrete components of larger mobilizations processes that occur over the course of conflict. Micromobilization uses micro-level data, whether at the level of the individual, geographic region, event, or phase in a conflict to better explain how mobilization occurs. The research points to the difficulty in developing a master motivation theory that accounts for motivations for joining resistance movements in all times and places. Instead, research on micromobilization processes seeks to identify how the motivations for mobilization changes according to the shifting dynamics within a conflict.[22]

Negotiated Settlements or Peace Agreements

Negotiated settlements or peace agreements are mutually agreed upon political arrangements that warring parties enter into to explicitly regulate or resolve the source of grievances fueling the conflict. Currently, more civil wars are ended through a negotiated settlement than a decisive military victory by either the incumbent government or rebels, a shift from patterns in conflict termination during the Cold War era. From 1940 until the end of the 1980s, most civil wars ended via a decisive military victory, whether by rebel forces or the incumbent government. From 1940 to 1999, civil wars were four times as likely to end via a military victory as a negotiated settlement.[23]

Nonviolent Resistance

Nonviolent resistance is a "socio-political action for applying power in a conflict without the use of violence." The techniques are outside the boundaries of conventional political processes, such as voting, lobbying and interest-group organizing. The persuasiveness of nonviolent campaigns derives from the continual tactical innovation that produces societal disruption. The tactics include boycotts, strikes, protests, sit-ins, stay-aways, and other forms of noncooperation and civil disobedience intended to pressure a ruling authority.[24]

Outbidding and Tactical Innovation

Outbidding is an action-counteraction dynamic between challengers and the state in which each side raises the stakes of engagement with one another. Tactical innovation by a resistance movement is the continual innovation and deployment of tactics designed to disrupt public order and mobilize supporters. Political violence often results from the dynamics of outbidding and tactical innovation as the state or resistance movements incrementally adopts more disruptive and violent tactics to disrupt or secure public order, respectively.[25]

Political Opportunity Set

The political opportunity set is comprised of three interrelated concepts that explain the political context surrounding conflict processes. The first, political opportunity structures, refers to the formal or permanent dimensions of the environment that shapes incentives for resistance. The second, the configuration of actors, looks at the existing relationships between powerful actors in the environment. The actors include a resistance group's potential allies, its adversaries, and influential bystanders. Lastly, the political opportunity set includes the dynamic process of ongoing interaction between resistance groups and their adversaries that impact the group's strategies, tactics, and political objectives or claims. Each of these concepts are powerful tools for better understanding resistance processes and their outcomes.[26]

Political Opportunity

The concept of political opportunity highlights factors in the political environment that incentivize individuals or groups to make decisions or take actions regarding participation in resistance that they otherwise might not have had those opportunities not been present. Political opportunities, conditions that make a regime more vulnerable to resistance, can include a decline in repression, divisions among the elite, or reforms that grant citizens greater political participation. While the conditions are important, the perceptions of these opportunities among individuals or groups are also significant. Political opportunity explains why some states might experience political resistance while other states do not, even though citizens in both have grievances.[27]

Political Outbidding

Political outbidding is a form of competition that arises between groups competing for scarce resources in the organizationally dense field of actors that often makes up a resistance movement. Groups and actors with similar goals often find themselves competing for sources of external funding, allies among the political elite, recruits, and legitimacy, among other resources. The competition encourages the groups to "outbid" one another to differentiate their group from the rest. One way to differentiate a group is through the use of extreme tactics, including violence, to represent themselves as the legitimate representative and protector of the community and their interests. As a result, more moderate groups that otherwise prefer nonviolence may have incentives to adopt more violent tactics to compete with the radicalized group.[28]

Political Process Model

The political process model describes resistance as the culmination of long-term political processes dictated by the existing power configurations in society. The model identifies three key factors in the political process model that explained the rise and decline of resistance movements, including the level of organization among the relevant population, positive assessment for the success of the insurgency, and the configuration of political actors within the government.[29]

Problem of Collective Action

The problem of collective action asserts that it is not rational for individuals to act on behalf of a group's interest because it produces public, not private, goods. Public goods are a class of goods that must be made available to everyone if they are made available at all. The collective action puzzle assumes that individuals are rational actors that make decisions and take actions based on calculations of self-interest. Rational actors prefer the decisions and actions that provide the greatest personal benefit or the lowest personal cost. As a result, individuals have an incentive to let others take on the burdensome task of resistance, or free-ride, because even nonparticipants will enjoy any benefits the group produces.[30]

Protest Cycle

A protest cycle is a period of heightened or intense mobilization across the social system. It is characterized by a diffusion of collective action; rapid innovation in strategies and tactics; a mixture of spontaneous and organized participation; and intense, repeated interactions between challengers of the status quo and state authorities.[31]

Radicalization

Radicalization occurs when members or leaders of a resistance movement shift toward ideological extremes or adopt increasingly disruptive or violent tactics. Radicalization does not necessarily cause decline, but it can serve as a mechanism for demobilization, particularly when one wing of the movement radicalizes in reaction to the moderation, co-optation, or institutionalization of a rival wing. The radicalized wing can lead to decreased legitimacy that hinders recruitment and operational relevance.[32]

Recidivism

Recidivism is a term used to describe the recurrence of civil war in a state that has experienced a period of peace or stability. The likelihood of recidivism is impacted by the type of conflict termination (negotiated settlement, military victory, or cease-fire agreement) and the details of the peace agreement, power-sharing arrangements, governing coalitions, elections, peacekeeping operations, and other post-conflict measures.[33]

Relative Deprivation

Relative deprivation describes the mismatch between peoples' levels of expectation regarding their economic situations and the realities of their economic situations. When relative deprivation occurs, individuals are more likely to participate in armed rebellion. Relative deprivation posits a relationship between people's perception of economic grievance, not the objective reality of an individual's economic situation. Relative deprivation is similar to horizontal inequality, except the latter theory focuses on groups, not individuals.[34]

Repertoires of Contention

Repertoires of contention are clustered acts resistance movements use to make claims against their targets, including state or occupational authorities. Each society, culture, historical movement, or other differentiated group has access to a set of stock contentious acts that are familiar and meaningful to the particular audience. The acts, because they are culturally embedded within a society, are familiar and resonate with the movement's audience. Repertoires of contention are described to mimic theatrical language, which captures how the acts are scripted but also offer room for improvisation such as "loosely scripted theater."[35]

Resource Endowments

Resource endowments are the configuration of resources that enable a resistance movement to overcame barriers to mobilization. The endowments are classified in two broad categories: economic endowments and social endowments. Economic endowments are finances derived from natural resources, taxing, criminal activity, or other sources. Social endowments are more ideational than financial, including shared beliefs, expectations, and norms among relevant groups. The endowments generate different membership profiles. Economic endowments tend to mobilize low-investment recruits that are attracted to personal financial gain. Resistance movements that rely on social endowments tend to mobilize high-commitment investor that favor activist rebellion over private gain. Researchers speculate that resistance movements that rely on economic endowments are more likely to wield indiscriminate violence against the local population, raising civilian casualty figures.[36]

Selective Incentives

Selective incentives are one explanation to the problem of collective action. Because individuals have incentives to free-ride, resistance leaders offer selective incentives or side payments to entice individuals to participate in collective action through the promise of personal reward. Possible selective incentives include land, money, loot, natural resources, and positions of authority that can attract mobilization in a wide variety of resistance activities, from strikes to violent rebellion. With the addition of selective incentives, participants receive multiple payoffs because they will arguably still also enjoy the public goods produced by the organization. This means that each individual benefits from private goods as well as public goods.[37]

Shadow Governments

Shadow governments are formal or informal nonstate organizations that strategically leverage governance activities to fulfill operational objectives relations to population support and control. The activities mimic the attributes or function of the state. Resistance movements use shadow governments to gain legitimacy among the local population, control the local population, undermine the state government, or extract valuable resources. Oftentimes, shadow governments will operate in tandem or in competition with formal state governance.[38]

Youth Bulge

A youth bulge is a demographic pattern in which a population has a disproportionately large youth population in comparison to the older population. A youth bulge can increase the risk a country experiences armed resistance when it is also accompanied by economic stagnation. Under these conditions, unemployment generates grievances and produces a large recruitment pool of military-age males.[39]

ENDNOTES

[1] Stacy Keogh, "The Survival of Religious Peace Movements: When Mobilization Increases as Political Opportunity Decreases," *Social Compass* 60, no. 4 (2013): 561–578; Traci M. Sawyers and David S. Meyers, "Missed Opportunities: Social Movement Abeyance and Public Policy," *Social Problems* 46, no. 2 (1999): 187–206; Verta Taylor, "Social Movement Continuity: The Women's Movement in Abeyance," *American Sociological Review* 54, no. 5 (1989): 761–775.

[2] Abraham Harold Maslow, "A Theory of Human Motivation," *Psychological Review* 50, no. 4 (1943): 370; Chuck Crossett and Jason Spitaletta, *Radicalization: Relevant Psychological and Sociological Concepts* (Ft. Meade, MD: Asymmetric Warfare Group, 2010); Aidan Kirby, "The London Bombers as 'Self-Starters': A Case Study in Indigenous Radicalization and the Emergence of Autonomous Cliques," *Studies in Conflict and Terrorism* 30, no. 5 (2007): 415–428; Ziad W. Munson, *The Making of Pro-life Activists: How Social Movement Mobilization Works* (Chicago: University of Chicago Press, 2010); Marc Sageman, *Understanding Terror Networks* (Philadelphia: University of Pennsylvania Press, 2004); Reinoud Leenders, "Collective Action and Mobilization in Dar'a: An Anatomy of the Onset of Syria's Popular Uprising," *Mobilization: An International Journal* 17, no. 4 (2012): 419–434.

[3] James Raymond Vreeland, "The Effect of Political Regime on Civil War Unpacking Anocracy," *Journal of Conflict Resolution* 52, no. 3 (2008): 401–425; Edward N. Muller, "Income Inequality, Regime Repressiveness, and Political Violence," *American Sociological Review* 50 (1985): 47–61; Patrick M. Regan and Sam R. Bell, "Changing Lanes or Stuck in

the Middle: Why are Anocracies More Prone to Civil War?" *Political Research Quarterly* 63, no. 4 (2010): 747–759; Polity Project, http://www.systemicpeace.org/polityproject.html.

4   Idean Salehyan and Kristian Skrede Gleditsch, "Refugees and the Spread of Civil War," *International Organization* 60, no. 2 (2006): 335–366; Peter Waldmann, "Is there a Culture of Violence in Colombia?" *Terrorism and Political Violence* 19 (2007): 593–609.

5   Kristian Skrede Gleditsch, "Transnational Dimensions of Civil War," *Journal of Peace Research* 44, no. 3 (2007): 293–309; Sarah Zukerman Daly, "Organizational Legacies of Violence: Conditions Favoring Insurgency Onset in Colombia, 1964-1984," *Journal of Peace Research* 49, no. 3 (2012): 473–491; J. Michael Quinn, T. David Mason and Mehmet Gurses, "Sustaining the Peace: Determinants of Civil War Recurrence," *International Interactions* 33 (2007): 167–193.

6   Stathis N. Kalyvas, "Micro-Level Studies of Violence in Civil War: Refining and Extending the Control-Collaboration Model," *Terrorism and Political Violence* 24, no. 4 (2012): 658–668; Stathis N. Kalyvas and Matthew Adam Kocher, "How 'Free' is Free Riding in Civil Wars?: Violence, Insurgency, and the Collective Action Problem," *World Politics* 59, no. 2 (2007): 177–216; Stathis N. Kalyvas, *The Logic of Violence in Civil War* (Cambridge: Cambridge University Press, 2006).

7   Jonathan Christiansen, "Four Stages of Social Movements," in *EBSCO Research Starters* (2009): 1–7; Frederick D. Miller, "The End of SDS and the Emergence of Weatherman: Demise through Success" in Jo Freeman and Victoria Johnson (eds.), *Waves of Protest: Social Movements Since the Sixties* (Lanham, MD: Rowman & Littlefield Publishers, 1999); Patrick G. Coy and Timothy Hedeen, "A Stage Model of Social Movement Co-Optation: Community Mediation in The United States," *The Sociological Quarterly* 46, no. 3 (2005): 405–435.

8   Donatella della Porta, "Competitive Escalation During Protest Cycles: Comparing Left-Wing and Religious Conflicts," in Demetriou, et. al, (eds.), *Dynamics of Political Violence: A Process Oriented Perspective on Radicalization and the Escalation of Political Conflict* (Surrey, England: Ashgate, 2014), 95.

9   David Keen, "The Economic Functions of Violence In Civil Wars," *The Adelphi Papers* 38, no. 320 (1998): 1–89; Alberto Abadie, "Poverty, Political Freedom, and The Roots of Terrorism," (working paper no. w10859, *National Bureau of Economic Research*, 2004); Léonce Ndikumana and Kisangani Emizet, "The Economics of Civil War: The Case of the Democratic Republic of Congo," in Paul Collier and Nicholas Sambanis (eds.) *Understanding Civil War: Evidence and Analysis, Volume I: Africa* (Washington, DC: World Bank, 2005).

10   John J. Macionis, *Sociology*, 9th ed. (Upper Saddle River, NJ: Prentice Hall, 2003), 619.

11   Andreas Wimmer, Lars-Erik Cederman, and Brian Min, "Ethnic Politics and Armed Conflict: A Configurational Analysis of a New Global Data Set," in *American Sociological Review* 74, no. 2 (2009): 316–337; Ernest Gellner, *Nations and Nationalism* (Ithaca and London: Cornell University Press, 2008); Ethnic Power Relations dataset, http://www.epr.ucla.edu/.

12   Jose G. Montalvo and Marta Reynal-Querol, "Ethnic Polarization and the Duration of Civil Wars," *Economics of Governance* 11, no. 2 (2010): 123–143; Marta Reynal-Querol, "Ethnicity, Political Systems, and Civil Wars," *Journal of Conflict Resolution* 46, no. 1 (2002): 29–54.

13   Sidney Tarrow, *Power in Movement: Social Movements and Contentious Politics* (Cambridge: Cambridge University Press, 1998), 54, 127, 189–190.

14   Other researchers have also combined geocoded inequality and ethnic settlement data. The measures of inequality include the use of satellite imagery of nightlight emissions which are highly correlated with economic activities. See Lars-Erik Cederman, Nils B.

Weidmann, and Nils-Christian Bormann, "Triangulating Horizontal Inequality: Toward Improved Conflict Analysis," *Journal of Peace Research* 52, no. 6 (2015): 806–821.

[15] Frances Stewart, ed., *Horizontal Inequalities and Conflict* (New York: Palgrave Macmillan, 2008); Lars Erik Cederman, Kristian Skrede Gleditsch, and Halvard Buhaug, *Inequality, Grievances, and Civil War* (New York: Cambridge University Press, 2013).

[16] Geocoded data are tied to specific geographic locations.

[17] Nathan Bos, ed., *Human Factors Considerations of Undergrounds in Insurgencies*, 2nd ed. (Fort Bragg, NC: USASOC, 2013), 123–130.

[18] Elisabeth Jean Wood, *Insurgent Collective Action and Civil War in El Salvador* (Cambridge, UK: Cambridge University Press, 2003).

[19] Sidney Tarrow, *Power in Movement: Social Movements and Contentious Politics* (Cambridge, UK: Cambridge University Press, 1998), 207–208; Rex D. Hopper, "The Revolutionary Process: A Frame of Reference for the Study of Revolutionary Movements," *Social Forces* 28, no. 3 (March 1950): 270–279.

[20] William Maley, "Building Legitimacy in Post-Taliban Afghanistan," in *State Building, Security and Social Change in Afghanistan* (San Francisco: Asia Foundation, 2008): 12, http://asiafoundation.org/resources/pdfs/2008surveycompanionvolumefinal.pdf.

[21] Kenneth E. Boulding, Conflict and Defense: A General Theory (New York: Harper, 1962); Halvard Buhaug, "Dude, Where's My Conflict?: LSG, Relative Strength, and the Location of Civil War," *Conflict Management and Peace Science* 20, no. 10 (2009): 1–22.

[22] Stathis N. Kalyvas, "Micro-Level Studies of Violence in Civil War: Refining and Extending the Control-Collaboration Model," *Terrorism and Political Violence* 24, no. 4 (2012): 658–668; Ana M. Arjona and Stathis N. Kalyvas, "Recruitment into Armed Groups in Colombia: A Survey of Demobilized Fighters," in Yvan Guichaoua (ed.), *Understanding Collective Political Violence* (London, UK: Palgrave McMillan, 2011); Stathis N. Kalyvas and Matthew Adam Kocher, "How 'Free' is Free Riding in civil wars?: Violence, Insurgency, and the Collective Action Problem," *World Politics* 59, no. 2 (2007): 177–216; Lorenzo Bosi and Donatella della Porta, "Micro-mobilization into Armed Groups: Ideological, Instrumental and Solidaristic Paths," *Qualitative Sociology* 35, no. 4 (2012): 361–383.

[23] Peter Wallenstein and Margareta Sollenberg, "Armed Conflicts, Conflict Termination and Peace Agreements, 1989-96," *Journal of Peace Research* 34, no. 3 (1997): 339–358; Gordon H. McCormick, Steven B. Horton, and Lauren A. Harrison, "Things Fall Apart: The Endgame Dynamics of Internal Wars," *Third World Quarterly* 28, no. 2 (2007): 321–367; Joakim Kreutz, "How and When Armed Conflicts End: Introducing the UCDP Conflict Termination Dataset," *Journal of Peace Research* 47, no. 2 (2010): 243–250; Monica Duffy Toft, *Securing the Peace: The Durable Settlement of Civil Wars* (Princeton: Princeton University Press, 2010); Matthew Hoddie and Caroline Hartzell, "Civil War Settlements and the Implementation of Military Power-Sharing Arrangements," *Journal of Peace Research* 40, no. 3 (2003): 303–320; Barbara F. Walter, "The Critical Barrier to Civil War Settlement," *International Organization* 51, no. 03 (1997): 335–364; Barbara F. Walter, "Bargaining Failures and Civil War," *Annual Review of Political Science* 12 (2009): 243–261.

[24] Erica Chenoweth and Maria J. Stephan, *Why Civil Resistance Works: The Strategic Logic of Nonviolent Conflict* (New York, NY: Columbia University Press, 2012), 12.

[25] Eitan Y. Alimi, Lorenzo Bosi, and Chares Demetriou, "Relational Dynamics and Processes of Radicalization: A Comparative Framework," *Mobilization: An International Journal* 17, no. 1 (2012): 7–26; Donatella della Porta, *Social Movements, Political Violence, and the State: A Comparative Analysis of Italy and Germany* (Cambridge: Cambridge University Press, 2006), 55–56; Doug McAdam, "Tactical Innovation and the Pace of Insurgency," *American Sociological Review* (1983): 735–754.

26 Hanspeter Kriesi, "Political Context and Opportunity," in David A. Snow, Sarah A. Soule, and Hanspeter Kriesi (eds.), *The Blackwell Companion to Social Movements* (Malden, MA: Blackwell Publishing, 2007), 67–90; Charles Tilly and Sidney Tarrow, *Contentious Politics* (New York, NY: Oxford University Press, 2007), 55–57.

27 Sidney Tarrow, *Power in Movement: Social Movements and Contentious Politics* (Cambridge, UK: Cambridge University Press, 1998); Doug McAdam, *Political Process and the Development of Black Insurgency, 1930–1970* (Chicago and London: University of Chicago Press, 1982); Charles Tilly and Sidney Tarrow, *Contentious Politics* (New York, NY: Oxford University Press, 2007).

28 Gianluca De Fazio, "Intramovement Competition and Political Outbidding as Mechanisms of Radicalization in Northern Ireland, 1968–1969," in Demetriou, et. al, (eds.), *Dynamics of Political Violence: A Process Oriented Perspective on Radicalization and the Escalation of Political Conflict* (Surrey, England: Ashgate, 2014), 117–120.

29 Doug McAdam, *Political Process and the Development of Black Insurgency, 1930–1970* (Chicago and London: University of Chicago Press, 1982), 40–42.

30 Mancur Olson, *The Logic of Collective Action* (Cambridge, MA: Harvard University Press, 1971).

31 Sidney Tarrow, *Power in Movement: Social Movements and Contentious Politics* (Cambridge, UK and New York, NY: Cambridge University Press, 1998), 142.

32 Sidney Tarrow, *Power in Movement: Social Movements and Contentious Politics* (Cambridge, UK: Cambridge University Press, 1998), 190, 207–208.

33 J. Michael Quinn, T. David Mason, and Mehmet Gurses, "Sustaining the Peace: Determinants of Civil War Recurrence," *International Interactions* 33, no. 2 (2007): 167–193; T. David Mason, Mehmet Gurses, Patrick T. Brandt, and Jason Michael Quinn, "When Civil Wars Recur: Conditions for Durable Peace after Civil Wars," *International Studies Perspectives* 12, no. 2 (2011): 186–187; Caroline Hartzell and Matthew Hoddie, "Institutionalizing Peace: Power Sharing and Post-Civil War Conflict Management," *American Journal of Political Science* 47, no. 2 (2003): 318–32; Karl Derouen, Jr., Jenna Lea, and Peter Wallensteen, "The Duration of Civil War Peace Agreements," *Conflict Management and Peace Science* 26, no. 4 (2009): 367–378.

34 Ted Robert Gurr, *Why Men Rebel* (London and New York: Routledge, 2011), 24.

35 Charles Tilly, *Popular Contention in Great Britain, 1758–1864* (Cambridge, MA: Harvard University Press, 1995), 41; Charles Tilly, *Regimes and Repertoires* (Chicago, IL: University of Chicago, 2010), 41; Sidney Tarrow, *Power in Movement: Social Movements and Contentious Politics* (Cambridge, UK and New York, NY: Cambridge University Press, 1998), 29–42.

36 Jeremy M. Weinstein, *Inside Rebellion: The Politics of Insurgent Violence* (Cambridge, UK: Cambridge University Press, 2007).

37 Mark Lichbach, *The Rebel's Dilemma* (Ann Arbor, MI: University of Michigan Press, 1998), 216; Samuel L. Popkin, *The Rational Peasant: The Political Economy of Rural Society in Vietnam* (Berkeley: University of California Press, 1979); Paul Collier and Anke Hoeffler, "Greed and Grievance in Civil War," *Oxford Economic Papers* 56 (2004): 563–595; Päivi Lujala, Nils Petter Gleditsch, and Elisabeth Gilmore, "A Diamond Curse? Civil War and a Lootable Resource," *Journal of Conflict Resolution* 49, no. 4 (2005): 538–562.

38 Nelson Kasfir, "Guerrillas and Civilian Participation: The National Resistance Army in Uganda, 1981–86," *Journal of Modern African Studies* 43, no. 2 (2005): 271–296; Zachariah Cherian Mampilly, *Rebel Rulers: Insurgent Governance and Civilian Life During War* (Ithaca: Cornell University Press, 2011); James B. Love, *Hezbollah: Social Services as a Source of Power* (Hurlbert Field, FL: JSOU Press, 2010); Summer Newton and Robert Leonhard, "Shadow

Government," in Robert Leonhard, (ed.), *Undergrounds in Insurgent, Revolutionary, and Resistance Warfare* (Fort Bragg, NC: USASOC, 2013), 131–168.

[39] Henrik Urdal, "A Clash of Generations? Youth Bulges and Political Violence," *International Studies Quarterly* 50, no. 3 (2006): 607–629; Nils Petter Gleditsch, Peter Wallensteen, Mikael Eriksson, Margareta Sollenberg, and Håvard Strand, "Armed Conflict 1946-2001: A New Dataset," *Journal of Peace Research* 39, no. 5 (2002): 615–6; Jack A. Goldstone, "Population and Security: How Demographic Change can Lead to Violent Conflict," *Journal of International Affairs* (2002): 3–21.

BIBLIOGRAPHY

Abadie, Alberto. "Poverty, Political Freedom, and The Roots of Terrorism." Working paper no. w10859. Cambridge, MA: National Bureau of Economic Research, 2004. http://www.nber.org/papers/w10859.

Addison, Tony and Syed Mansoob Murshed. *The Fiscal Dimensions of Conflict and Reconstruction.* Helsinki: United Nations University, World Institute for Development Economics Research, 2001.

Agan, Summer D., ed. *ARIS Narratives and Competing Messages.* Fort Bragg, NC: USASOC, forthcoming.

Alberts, David S. and Richard E. Hayes. *Power to the Edge: Command and Control in the Information Age.* CCRP Publication Series, 2003. Kindle.

Albrecht, Holger. "Does Coup-proofing Work? Political–Military Relations in Authoritarian Regimes Amid the Arab Uprisings." *Mediterranean Politics* 20, no. 1 (2015): 36-54.

Alimi, Eitan Y., Lorenzo Bosi, and Chares Demetriou. "Relational Dynamics and Processes of Radicalization: A Comparative Framework." *Mobilization: An International Journal* 17, no. 1 (2012): 7–26.

Amery, Julien. "Of Resistance," in *The Nineteenth Century Magazine* (March 1949): 138–149.

Ardrey, Robert. *The Territorial Imperative: A Personal Inquiry into the Animal Origins of Property and Nations.* New York: Atheneum, 1966.

Arjona, Ana M. and Stathis N. Kalyvas. "Recruitment into Armed Groups in Colombia: A Survey of Demobilized Fighters." In *Understanding Collective Political Violence,* edited by Yvan Guichaoua, 143–174. London: Palgrave McMillan, 2011.

Barth, Frederik. "Introduction." In *Ethnic Groups and Boundaries,* edited by Frederik Barth, 9–38. Long Grove, IL: Waveland Press, 1998.

Bellin, Eva. "Reconsidering the Robustness of Authoritarianism in the Middle East: Lessons from the Arab Spring." *Comparative Politics* 44, no. 2 (2012): 127–149.

———. "The Robustness of Authoritarianism in the Middle East: Exceptionalism in Comparative Perspective." *Comparative Politics* 36, no. 2 (2004): 139–157.

Benford, Robert D. and David A. Snow, "Framing Processes and Social Movements: An Overview and Assessment." *Annual Review of Sociology* 26 (2000): 614.

Bennett, W. Lance and Alexandra Segerburg. *The Logic of Connective Action: Digital Media and the Personalizatin of Contentious Politics.* Cambridge, UK: Cambrige University Press, 2013.

Binchy, D.A. "A Pre-Christian Survival in Mediaeval Irish Hagiography." In *Ireland in Medieval Europe: Studies in Memory of Kathleen Hughes*, edited by Dorothy Whitelock, 165–178. Cambridge, UK: Cambridge University Press, 1982.

Bos, Nathan, ed. *Human Factors Considerations of Undergrounds in Insurgencies*, 2nd ed. Fort Bragg, NC: USASOC, 2013.

Bosi, Lorenzo. "Safe Territories and Violent Political Organizations." *Nationalism and Ethnic Politics* 19, no. 1 (2013): 80–101.

Bosi, Lorenzo, Chares Demetriou, and Stefan Malthaner. "A Contentious Politics Approach to the Explanation of Radicalization." In *Dynamics of Political Violence: A Process Oriented Perspective on Radicalization and the Escalation of Political Conflict*, edited by Lorenzo Bosi, Chares Demetriou, and Stefan Malthaner, 1–25. Surrey, England: Ashgate, 2014.

Boulding, Kenneth E. *Conflict and Defense: A General Theory.* New York: Harper, 1962.

Bowyer Bell, J. "Dragonworld (II): Deception, Tradecraft, and the Provisional IRA." *International Journal of Intelligence and Counterintelligence* 8, no. 1 (1995): 21–50.

Braithwaite, Alex. "Does Poverty Cause Conflict?: Isolating the Causal Origins of the Conflict Trap." *Conflict Management and Peace Science* 33, no. 1 (2016): 45–66.

Brinton, Crane. *The Anatomy of Revolution.* Revised Edition. New York, NY: Vintage Books, 1965.

Brubaker, Rogers. *Ethnicity Without Groups.* Cambridge, MA: Harvard University Press, 2004.

Bryant, Raymond L. "Shifting the Cultivator: The Politics of Teak Regeneration in Colonial Burma." *Modern Asian Studies* 28, no. 2 (1994): 225–250.

Buhaug, Halvard. "Dude, Where's My Conflict?: LSG, Relative Strength, and the Location of Civil War." *Conflict Management and Peace Science* 20, no. 10 (2009): 1–22.

Buhaug, Halvard and Jan Ketil Rød. "Local Determinants of African Civil Wars, 1970–2001." *Political Geography* 25, no. 3 (2006): 315–335.

Buhaug, Halvard and Scott Gates. "The Geography of Civil War." *Journal of Peace Research* 39, no.4 (2002): 417–433.

Buhaug, Halvard, Tor A. Benaminsen, Espen Sjaastad, and Ole Magnus Theisen. "Climate Variability, Food Production Shocks, and Violent Conflict in Sub-Saharan Africa." Environmental Research Letters 10, no. 12 (2015): 125015.

Buikema, Ron and Matt Burger. "Sendoro Luminoso (Shining Path)." In ARIS Casebook on Insurgency and Revolutionary Warfare Volume II: 1962-2009, edited by Chuck Crossett, 55-86. Fort Bragg, NC: USASOC, 2012.

———. "Fuerzas Armadas Revolucionarias De Colombia (FARC)," in ARIS Casebook on Insurgency and Revolutionary Warfare, edited by Chuck Crossett, 31-54. Fort Bragg, NC: USASOC, 2013.

Butcher, Charles. "'Capital Punishment': Bargaining and the Geography of Civil War." *Journal of Peace Research* 52, no. 2 (2015): 171–186.

Carey, Sabine C. "The Dynamic Relationship between Protest and Repression." *Political Research Quarterly* 59, no. 1 (2006): 1–11.

———. "The Use of Repression as a Response to Domestic Dissent." *Political Studies* 58, no. 1 (2010): 167–186.

Carr, C. Lynn. "Tomboy Resistance and Conformity: Agency in Social Psychological Gender Theory." *Gender & Society* 12 (1998): 528–553.

Cederman, Lars-Erik, Andreas Wimmer, and Brian Min. "Why do Ethnic Groups Rebel? New Data and Analysis." *World Politics* 62, no. 01 (2010): 87–119.

Cederman, Lars-Erik, Kristian Skrede Gleditsch, and Halvard Buhaug. *Inequality, Grievances, and Civil War.* New York: Cambridge University Press, 2013.

Cederman, Lars-Erik and Luc Girardin. "Beyond Fractionalization: Mapping Ethnicity onto Nationalist Insurgencies." *American Political Science Review* 101, no. 01 (2007): 173–185.

Cederman, Lars-Erik, Nils B. Weidmann, and Nils-Christian Bormann. "Triangulating Horizontal Inequality: Toward Improved Conflict Analysis." *Journal of Peace Research* 52, no. 6 (2015): 806–821.

Central Intelligence Agency. *Guide to the Analysis of Insurgency.* United States: Author, 2012. https://www.hsdl.org/?view&did=713599.

Chenoweth, Erica and Maria J. Stephan. *Why Civil Resistance Works: The Strategic Logic of Nonviolent Conflict.* New York: Columbia University Press, 2011.

Chenoweth, Erica and Orion A. Lewis. "Unpacking Nonviolent Campaigns: Introducing the NAVCO 2.0 Dataset." *Journal of Peace Research* 50, no. 3 (2013): 415–423.

Christiansen, Jonathan. "Social Movements & Collective Behavior: Four Stages of Social Movements." *Research Starters Academic Topic Overviews.* Ipswich, MA: EBSCO, 2009. http://fliphtml5.com/hqig/sewz/basic.

Collier, Paul and Anke Hoeffler. "Greed and Grievance in Civil War." *Oxford Economic Papers* 56, no. 4 (2004): 563–595.

Collier, Paul, Anke Hoeffler, and Måns Söderbom. "On the Duration of Civil War." *Journal of Peace Research* 41, no. 3 (2004): 253–273.

Collier, Paul and Nicholas Sambanis. "Understanding Civil War: A New Agenda." *Journal of Conflict Resolution* 46, no. 1 (2002): 3–12.

Condit, D. M. *Case Study in Guerrilla War: Greece during World War II,* edited by Erin M. Richardson. Fort Bragg, NC: USASOC, 2014.

Conley, Jerry. "Movement for the Emancipation of the Niger Delta (MEND)," in *Casebook on Insurgency and Revolutionary Warfare Volume II: 1962–2009,* edited by Chuck Crossett, 567–592. Fort Bragg, NC: USASOC, 2012.

Crosset, Chuck ed. *Casebook on Insurgency and Revolutionary Warfare, Volume II: 1962–2009.* Fort Bragg, NC: USASOC, 2012.

Crossett, Chuck and Summer Newton. "The Provisional Irish Republican Army (PIRA): 1969-2001." In *Casebook on Insurgency and Revolutionary Warfare Volume II: 1962–2009,* edited by Chuck Crossett, 379–422. Fort Bragg, NC: USASOC, 2012.

———. "Solidarity." In *Casebook on Insurgency and Revolutionary Warfare Volume II: 1962–2009,* edited by Chuck Crossett, 645–672. Fort Bragg, NC: USASOC, 2013.

Crossett, Chuck and Jason Spitaletta. *Radicalization: Relevant Psychological and Sociological Concepts.* Ft. Meade, MD: Asymmetric Warfare Group, 2010.

Cunningham, David E. "Blocking Resolution: How External States can Prolong Civil Wars." *Journal of Peace Research* 47, no. 2 (2010): 115–127.

Cunningham, David E., Kristian Skrede Gleditsch, and Idean Salehyan. "It Takes Two: A Dyadic Analysis of Civil War Duration and Outcome." *The Journal of Conflict Resolution* 53, no. 4 (2009): 570–597.

Daly, Sarah Zukerman. "Organizational Legacies of Violence: Conditions Favoring Insurgency Onset in Colombia, 1964–1984." *Journal of Peace Research* 49, no. 3 (2012): 473–491.

Dassel, Kurt. "Civilians, Soldiers, and Strife: Domestic Sources of International Aggression." *International Security* 23, no. 1 (1998): 107–140.

Davenport, Christian. *How Social Movements Die: Repression and Demobilization of the Republic of New Africa.* New York: Cambridge University Press, 2015.

Davenport, Christian and Will H. Moore. "The Arab Spring, Winter, and Back Again?(Re) Introducing the Dissent-Repression Nexus with a Twist." *International Interactions* 38, no. 5 (2012): 704–713.

De Fazio, Gianluca. "Intramovement Competition and Political Outbidding as Mechanisms of Radicalization in Northern Ireland, 1968-1969." In *Dynamics of Political Violence: A Process Oriented Perspective on Radicalization and the Escalation of Political Conflict*, edited by Lorenzo Bosi, Chares Demetriou, and Stefan Malthaner, 115–13. Surrey, England: Ashgate, 2014.

de la Calle, Luis and Ignacio Sánchez-Cuenca. "How Armed Groups Fight: Territorial Control and Violent Tactics." *Studies in Conflict & Terrorism* 38, Vol. 10, 796–797.

———. "In Search of the Core of Terrorism." *International Studies Review* 14 (2012): 475–497.

della Porta, Donatella. *Social Movements, Political Violence, and the State: A Comparative Analysis of Italy and Germany.* Cambridge, UK: Cambridge University Press, 2006.

———. "Competitive Escalation During Protest Cycles: Comparing Left-Wing and Religious Conflicts." In *Dynamics of Political Violence: A Process Oriented Perspective on Radicalization and the Escalation of Political Conflict*, edited by Lorenzo Bosi, Chares Demetriou, and Stefan Malthaner, 93–114. Surrey, England: Ashgate, 2014.

Denny, Elaine K. and Barbara F. Walter. "Ethnicity and Civil War." *Journal of Peace Research* 51, no. 2 (2014): 199–212.

DeRouen, Karl R., Jr. and David Sobek. "The Dynamics of Civil War Duration and Outcome." *Journal of Peace Research* 41, no. 3 (2004): 303–320.

DeRouen, Karl R., Jr., Jenna Lea, and Peter Wallensteen. "The Duration of Civil War Peace Agreements." *Conflict Management and Peace Science* 26, no. 4 (2009): 367–387.

Diani, Mario. "The Concept of Social Movement." *The Sociological Review* 40, no. 1 (1992): 1–25.

Edwards, Lyford P. *The Natural History of Revolutions.* Chicago, IL: University of Chicago Press, 1927.

Ellingsen, Tanja. "Colorful Community or Ethnic Witches' Brew? Multiethnicity and Domestic Conflict During and After the Cold War." *Journal of Conflict Resolution* 44, no. 2 (2000): 228–249.

Euben, Roxanne L. "Humiliation and the Political Mobilization of Masculinity." *Political Theory* (2015): 1–33.

Fearon, James D. "Governance and Civil War Onset" *World Development Report 2011 Background Paper,* 2011: https://openknowledge.worldbank.org/bitstream/handle/10986/9123/WDR2011_0002.pdf.

———. "Iraq's Civil War." *Foreign Affairs,* March/April 2007, http://www.foreignaffairs.com/articles/62443/james-d-fearon/iraqs-civil-war.

———. "Why Do Some Civil Wars Last So Much Longer than Others?" *Journal of Peace Research* 41, no. 3(2004): 275–301.

Fearon, James D. and David D. Laitin. "Civil War Termination." Draft paper, Annual Meetings of the American Political Science Association, September 12, 2008.

———. "Ethnicity, Insurgency, and Civil War." *American Political Science Review* 97, no. 1 (2003): 75–90.

Fields, Rona M. *Martyrdom: The Psychology, Theology, and Politics of Self-Sacrifice.* New York: Praeger, 2004.

Flavin, William. "Planning for Conflict Termination and Post-Conflict Success." *Parameters* 33, no. 3 (Autumn 2003): 95–112.

Florea, Adrian. "Where Do We Go from Here? Conceptual, Theoretical, and Methodological Gaps in the Large-n Civil War Research Program." *International Studies Review* 14, no. 1 (2012): 78–98.

Fortna, Virginia Page. "Does Peacekeeping Keep Peace? International Intervention and the Duration of Peace after Civil War." *International Studies Quarterly* 48, no, 2 (2004): 269–292.

Gamson, William A. *The Strategy of Social Protest.* Homewood, IL: Dorsey Press, 1975.

———. "Constructing Social Protest." In S*ocial Movements and Culture,* edited by Hank Johnston and Bert Klandermas. Minneapolis: University of Minnesota Press, 1995.

Garton Ash, Timothy. *The Polish Revolution: Solidarity.* New Haven, CT: Yale University Press, 1999.

Gellner, Ernest. *Nations and Nationalism.* Ithaca, NY and London, UK: Cornell University Press, 2008.

Gent, Stephen E. "Going in When It Counts: Military Intervention and the Outcome of Civil Conflicts." *International Studies Quarterly* 52, no. 4 (2008): 713–735.

Gleditsch, Kristian Skrede. "Transnational Dimensions of Civil War." *Journal of Peace Research* 44, no. 3 (2007): 293–309.

Gleditsch, Kristian Skrede, Nils Petter, Peter Wallensteen, Mikael Eriksson, Margareta Sollenberg, and Håvard Strand. "Armed Conflict 1946-2001: A New Dataset." *Journal of Peace Research* 39, no. 5 (2002): 615–637.

"Global Health Observatory (GHO) Data: Urban Population Growth." *World Health Organization.* Accessed February 3, 2014, http://www.who.int/gho/urban_health/situation_trends/urban_population_growth_text/en/.

Goffman, Erving. *Frame Analysis: An Essay on the Organization of Experience.* New York: Harper Colophon, 1974.

Goldstone, Jack A. "Population and Security: How Demographic Change can Lead to Violent Conflict." *Journal of International Affairs* (2002): 3–21.

———. *Revolution and Rebellion in the Early Modern World.* Berkeley, CA and Los Angeles, CA: University of California Press, 1991.

Goldstone, Jack A., Robert H. Bates, David L. Epstein, Ted Robert Gurr, Michael B. Lustik, Monty G. Marshall, Jay Ulfelder, and Mark Woodward. "A Global Model for Forecasting Political Instability," *American Journal of Political Science* 54, no. 1 (January 2010): 190–208.

Goodwin, Jeff, James M. Jasper, and Francesca Polletta. "Emotional Dimensions of Social Movements." In *The Blackwell Companion to Social Movements*, edited by David A. Snow, Sarah A. Soule, and Hanspeter Kriesi, 413–432. Malden, MA: Blackwell Publishing, 2007.

Goodwin, Jeff, and Steven Paff. "Emotion Work in High-Risk Social Movements." In *Passionate Politics: Emotions and Social Movements*, edited by Jeff Goodwin, James M. Jasper, and Francesca Polletta, 611–635. Chicago, IL: University of Chicago Press, 2001.

Gregg, Nina. "'Trying to put First Things First': Negotiating Subjectives in a Workplace Organizing Campaign." In *Negotiating at the Margins: The Gendered Discourses of Power and Resistance*, edited by Sue Fisher and Kathy Davis, 172–204. New Brunswick, NJ: Rutgers University Press, 1993.

Gurr, Ted Robert. Why Men Rebel. London, UK and New York, NY: Routledge, 2011.

Hartzell, Caroline and Matthew Hoddie. "Institutionalizing Peace: Power Sharing and Post-Civil War Conflict Management." *American Journal of Political Science* 47, no. 2 (2003): 318–332.

Hartzell, Caroline, Matthew Hoddie and Donald Rothchild. "Stabilizing the Peace after Civil War: An Investigation of Some Key Variables." *International Organization* 55, no. 1 (2001): 183–208.

Hazen, Jennifer M. *What Rebels Want: Resources and Supply Networks in Wartime*. Ithaca: Cornell University Press, 2013.

Heger, Lindsay, Danielle Jung, and Wendy H. Wong. "Organizing for Resistance: How Group Structure Impacts the Character of Violence." *Terrorism and Political Violence* 24, no. 5 (2012): 743–768.

Heger, Lindsay and Idean Salehyan. "Ruthless Rulers: Coalition Size and the Severity of Civil Conflict." *International Studies Quarterly* 51, no. 2 (2007): 385–403.

Hegre, Håvard. "Democracy and Armed Conflict." *Journal of Peace Research* 51, no. 2 (2014): 159–172.

Hegre, Håvard and Nicholas Sambanis. "Sensitivity Analysis of Empirical Results on Civil War Onset." *Journal of Conflict Resolution* 50, no. 4 (2006): 508–535.

Hegre, Håvard and Tanja Ellingsen, Scott Gates and Nils Petter Gleditsch. "Toward a Democratic Civil Peace? Democracy, Political Change, and Civil War, 1816-1992." *The American Political Science Review* 95, no. 1 (2001): 33–48.

Heidelberg Institute for International Conflict Research. *Conflict Barometer 2010* no. 19. Heidelberg, Germany: Author, 2010, http://hiik.de/en/konfliktbarometer/pdf/ConflictBarometer_2010.pdf.

———. *Conflict Barometer 2013* no. 22. 2014, http://hiik.de/de/downloads/data/downloads_2013/ConflictBarometer2013.pdf.

Hironaka, Ann. Neverending Wars: *The International Community, Weak States, and the Perpetuation of Civil War.* Cambridge, MA: Harvard University Press, 2005.

Hoddie, Matthew and Caroline Hartzell. "Civil War Settlements and the Implementation of Military Power-Sharing Arrangements." *Journal of Peace Research* 40, no. 3 (2003): 303–320.

Hollander, Jocelyn A. and Rachel L. Einwohner. "Conceptualizing Resistance." *Sociological Forum* 19, no. 4 (2004): 533–554.

Hopper, Rex D. "The Revolutionary Process: A Frame of Reference for the Study of Revolutionary Movements." *Social Forces* 28, no. 3 (1950 March): 270–279.

Horowitz, Donald L. *Ethnic Groups in Conflict.* Berkeley: University of California Press, 1985.

Humphreys, Macartan and Jeremy M. Weinstein. "Who Fights? The Determinants of Participation in Civil War." *American Journal of Political Science* 52, no. 2 (2008): 436–455.

"International Country Risk Guide (ICRG)." *The PRS Group.* Last modified 2016. https://www.prsgroup.com/about-us/our-two-methodologies/icrg.

Jackson, Maurice, Eleanora Petersen, James Bull, Sverre Monsen, and Patricia Richmond. "The Failure of an Incipient Social Movement." *The Pacific Sociological Review* 3, no. 1 (Spring, 1960): 35–40.

Jansen, Johannes J. G. *The Neglected Duty: The Creed of Sadat's Assassins and Islamic Resurgence in the Middle East.* New York, NY: Macmillan, 1986.

Jasper, James M. *The Art of Moral Protest.* Chicago, IL: University of Chicago Press, 1997.

Jasper, James M. and Jane Poulsen. "Recruiting Strangers and Friends: Moral Shocks and Social Networks in Animal Rights Activism and Anti-Nuclear Protests." *Social Problems* 42(1995): 493–512.

Johnston, Hank. "The Mechanism of Emotion in Violent Protest." In *The Dynamics of Political Violence: A Process-Oriented Perspective on Radicalization and the Escalation of Political Conflict*, edited by Charles Demetriou, Stefan Malthaner, Lorenzo Bosi, 27–50. Burlington, VT: Ashgate, 2014.

Joshi, Madhav and T. David Mason. "Civil War Settlements, Size of Governing Coalition and Durability of Peace in the Post-Civil War States." Paper presented at the International Studies Association Annual Convention, New Orleans, LA, February 2010.

Kalyvas, Stathis N. *The Logic of Violence in Civil War.* New York, NY: Cambridge University Press, 2006.

———. "Promises and Pitfalls of an Emerging Research Program: The Microdynamics of Civil War." In *Order, Conflict, Violence*, edited by Stathis N. Kalyvas, Ian Shapiro and Tarek Masoud, 397–421. Cambridge, UK: Cambridge University Press, 2008.

———. "Micro-Level Studies of Violence in Civil War: Refining and Extending the Control-Collaboration Mode." *Terrorism and Political Violence* 24, no. 4(2012): 658–668.

Kalyvas, Stathis N. and Laia Balcells. "International System and Technologies of Rebellion: How the End of the Cold War Shaped Internal Conflict." *American Political Science Review* 104, no. 3 (2010): 415–429.

Kalyvas, Stathis N. and Matthew Adam Kocher. "How 'Free' is Free Riding in Civil Wars? Violence, Insurgency, and the Collective Action Problem." *World Politics* 59, no. 2(2007): 177–216.

———. "The Dynamics of Violence in Vietnam: An Analysis of the Hamlet Evaluation System (HES)." *Journal of Peace Research* 46, no. 3(2009): 335–355.

Kasfir, Nelson. "Guerrillas and Civilian Participation: The National Resistance Army in Uganda, 1981–86." *Journal of Modern African Studies* 43, no. 2 (2005): 272.

Keen, David. "The Economic Functions Of Violence In Civil Wars." *The Adelphi Papers* 38, no. 320 (1998): 1–89.

Keogh, Stacy. "The Survival of Religious Peace Movements: When Mobilization Increases as Political Opportunity Decreases." *Social Compass* 60, no. 4 (December, 2013): 561–578.

Khawaja, Marwan. "Repression and Popular Collective Action: Evidence from the West Bank." *Sociological Forum* 8, no. 1 (1993): 47–71.

Kirby, Aidan. "The London Bombers as 'Self-Starters': A Case Study in Indigenous Radicalization and the Emergence of Autonomous Cliques." *Studies in Conflict and Terrorism* 30, no. 5 (2007): 415–428.

Kleinberg, Jon. "The Convergence of Social and Technological Networks." *Communications of the ACM* 53, no. 11 (2008): 66–72.

Koopmans, Ruud. "Dynamics of Repression and Mobilization: The German Extreme Right in the 1990s." *Mobilization: An International Journal* 2, no. 2 (1997): 149–164.

———. "Protest in Time and Space: The Evolution of Waves of Contention." In *The Blackwell Companion to Social Movements*, edited by David A. Snow, Sarah A. Soule, and Hanspeter Kriesi, 19–46. Malden, MA: Blackwell Publishing, 2004.

Kreutz, Joakim. "How and When Armed Conflicts End: Introducing the UCDP Conflict Termination Dataset." *Journal of Peace Research* 47, no. 2 (2010): 243–250.

Kriesi, Hanspeter. "Political Context and Opportunity." In *The Blackwell Companion to Social Movements*, edited by David A. Snow, Sarah A. Soule, and Hanspeter Kriesi, 67–90. Malden, MA: Blackwell Publishing, 2007.

Kubota, Yuichi. *Armed Groups in Cambodian Civil War: The Stronghold and Beyond*. New York, NY: Palgrave Macmillan 2013.

Lacquer, Walter. *A History of Terrorism*. New Brunswick, NJ: Transaction Publishers, 2012.

Lauber, W. Sam (editor). *ARIS Understanding the States of Resistance*. Fort Bragg, NC: USASOC, forthcoming.

Lawrence, Adria and Erica Chenoweth. "Introduction." In *Rethinking Violence: States and Non-State Actors in Conflict*, edited by Erica Chenoweth and Adria Lawrence, 1–20. Cambridge, MA: The MIT Press, 2010.

Lawson, George. "Revolution, Nonviolence, and the Arab Uprisings." *Mobilization: An International Quarterly* 20, no. 4 (2015): 453–470.

Leenders, Reinoud. "Collective Action and Mobilization in Dar'a: An Anatomy of the Onset of Syria's Popular Uprising." *Mobilization: An International Journal* 17, no. 4(2012): 419–434.

Licklider, Roy. "The Consequences of Negotiated Settlements in Civil Wars, 1945-1993." *American Political Science Review* 89, no. 3 (1995): 681–690.

Lichbach, Mark Irving. "Deterrence or Escalation? The Puzzle of Aggregate Studies of Repression and Dissent." *Journal of Conflict Resolution* 31, no. 2 (1987): 266–297.

———. *The Rebel's Dilemma.* Ann Arbor, MI: University of Michigan Press, 1998.

Lichbach, Mark Irving and Ted Robert Gurr. "The Conflict Process: A Formal Model." *The Journal of Conflict Resolution* 25, no. 1 (1981): 3–29.

Leonhard, Robert, ed. *ARIS Undergrounds in Insurgent, Revolutionary, and Resistance Warfare.* Second edition. Fort Bragg, NC: USASOC, 2013.

Love, James. B. *Hezbollah: Social Services as a Source of Power.* Hurlbert Field, FL: JSOU Press, 2010.

Lujala, Päivi. "The Spoils of Nature: Armed Civil Conflict and Rebel Access to Natural Resources." *Journal of Peace Research* 47, no. 1(2010): 15–28.

Lujala, Päivi and Nils Petter Gleditsch, and Elisabeth Gilmore. "A Diamond Curse? Civil War and a Lootable Resource." *Journal of Conflict Resolution* 49, no. 4 (2005): 538–562.

Lyons, H. A., and H. J. Harbinson. "A Comparison of Political and Non-Political Murderers in Northern Ireland, 1974–84." *Medicine, Science and the Law* 26, no. 3 (1986): 193–198.

Macionis, John J. *Sociology,* Ninth Edition. Upper Saddle River, NJ: Prentice Hall, 2003.

Madsen, Elizabeth Leahy, Beatrice Daumerie, and Karen Hardee. "The Effects of Age Structure on Development: Policy and Issue Brief." Washington DC: Population Alliance Initiative, 2010. Brief. http://pai.org/wp-content/uploads/2012/01/SOTC_PIB.pdf.

Mampilly, Zachariah Cherian. *Rebel Rulers: Insurgent Governance and Civilian Life during War.* Ithaca, NY: Cornell University Press, 2011.

Marshall, Monty G., Ted Robert Gurr, and Barbara Harff. "PITF Problem Set Codebook." *Center for Global Policy*, April, 2009. Accessed at http://globalpolicy.gmu.edu/political-instability-task-force-home/pitf-problem-set-codebook/.

Maslow, Abraham Harold. "A Theory of Human Motivation." *Psychological Review* 50, no. 4 (1943): 370.

Mason, David T. and Dale A. Krane. "The Political Economy of Death Squads: Toward a Theory of the Impact of State-sanctioned Terror." *International Studies Quarterly* 33, no. 2 (1989): 175–198.

Mason, T. David, Mehmet Gurses, Patrick T. Brandt, and Jason Michael Quinn. "When Civil Wars Recur: Conditions for Durable Peace after Civil Wars." *International Studies Perspectives* 12, no. 2 (2011): 171–189.

McAdam, Doug. "Tactical Innovation and the Pace of Insurgency." *American Sociological Review* (1983): 735–754.

McCauley, Clark R. and Sophia Moskalenko, *Friction: How Radicalization Happens to Them and Us.* New York: Oxford University Press, 2011.

McCormick, Gordon H., Steven B. Horton, and Lauren A. Harrison. "Things Fall Apart: The Endgame Dynamics of Internal Wars." *Third World Quarterly* 28, no. 2 (2007): 321–367.

Meadows, Paul. "Sequence in Revolution." *American Sociological Review* 6, no. 5 (October, 1941): 702–709.

Meernik, James D., Angela Nichols and Kimi L. King. "The Impact of International Tribunals and Domestic Trials on Peace and Human Rights After Civil War." *International Studies Perspectives* 11, no. 4 (2010): 309–334.

Meyer, David S. "Protest and Political Opportunities." *Annual Review of Sociology* 30 (2004): 125–145.

Miguel, Edward, Shanker Satyanath, and Ernest Sergenti. "Economic Shocks and Civil Conflict: An Instrumental Variables Approach." *Journal of Political Economy* 112, no. 4 (2004): 725–753.

Miller, Frederick D. "The End of SDS and the Emergence of Weatherman: Demise through Success." In *Waves of Protest: Social Movements Since the Sixties,* edited by Jo Freeman and Victoria Johnson, 303–324. Lanham, MD: Rowman & Littlefield Publishers, 1999.

Mirzaei, Sanaz. "Palestinian Liberation Organization (PLO): 1964–2009." In *Casebook on Insurgency and Revolutionary Warfare Volume II: 1962–2009*, editor Chuck Crossett, 213–236. Fort Bragg, NC: United States Army Special Operations Command, 2012.

Modigliani, Andre and Francois Rochat. "The Role of Interaction Sequences and the Timing of Resistance in Shaping Obedience and Defiance to Authority." *Journal of Social Issues* 51 (1995): 107–123.

Moghadam, Assaf. "Mayhem, Myths, and Martyrdom: The Shi'a Conception of Jihad." *Terrorism and Political Violence* 19, no. 1 (2007): 125–143.

Moller, Herbert. "Youth as a Force in the Modern World." *Comparative Studies in Society and History* 10, no. 03 (1968): 237–260.

Molner, Andrew R., William A. Lybrand, Lorna Hahn, James L. Kirkman, and Peter B. Riddleberger. *Undergrounds in Insurgent, Revolutionary, and Resistance Warfare*. Washington DC: Special Operations Research Office, 1963.

Montalvo, Jose G. and Marta Reynal-Querol. "Ethnic Polarization and the Duration of Civil Wars." *Economics of Governance* 11, no. 2 (2010): 123–143.

Moore, Will H. "Action-Reaction or Rational Expectations? Reciprocity and the Domestic-International Conflict Nexus during the 'Rhodesia Problem.'" *Journal of Conflict Resolution* 39, no. 1 (1995): 129–167.

Muller, Edward N. "Income Inequality, Regime Repressiveness, and Political Violence." *American Sociological Review* 50 (1985): 47–61.

Ndikumana, Léonce and Kisangani Emizet. "The Economics of Civil War: The Case of the Democratic Republic of Congo." In *Understanding Civil War: Evidence and Analysis, Volume I: Africa*, edited by Paul Collier and Nicholas Sambanis, 63–87. Washington DC: World Bank, 2005.

Newton, Summer D. and Robert Leonhard. "Shadow Government." In *Undergrounds in Insurgent, Revolutionary, and Resistance Warfare*, edited by Robert Leonhard, 131–168. Fort Bragg, NC: USASOC, 2013.

Nisbett, Richard E. and Dov Cohen. *Culture of Honor: The Psychology of Violence in the South*. Boulder, CO: Westview Press, 1996.

Nordas, Ranghild and Christian Davenport. "Fight the Youth: Youth Bulges and State Repression." *American Journal of Political Science* 57, no. 4 (2013): 926–940.

O'Brien, Kevin J. "Rightful Resistance." *World Politics* 49, no. 1 (1996): 31–55.

O'Connor, Frances Patrick. "Armed Social Movements and Insurgency: The PKK and its Communities of Support." PhD diss., European University Institute, 2014. https://www.eui.eu/Documents/DepartmentsCentres/SPS/ThesesDefended2014/OConnorbioandabstract.pdf.

O'Connor, Frances Patrick and Leonidas Oikonomakis. "Preconflict Mobilization Strategies and Urban-Rural Transition: The Case of the PKK and the FLN/EZLN." *Mobilization: An International Quarterly* 20, no. 3(2015): 379–399.

Olson, Mancur. *The Logic of Collective Action.* Cambridge, MA: Harvard University Press, 1971.

Osa, Maryjane. "Networks in Opposition: Linking Organizations through Activists in the Polish People's Republic." In *Social Movements and Networks: Relational Approaches to Collective Action,* edited by Mario Diani and Doug McAdam, 77–105. Oxford, UK: Oxford University Press, 2003.

Ostby, Gudrun. "Polarization, Horizontal Inequalities and Violent Civil Conflict." *Journal of Peace Research* 45, no. 2 (2008): 143–162.

Ostby, Gudrun, Henrik Urdal, Mohammad Zulfan Tadjoeddin, S. Mansoob Murshed, and Havard Strand. "Population Pressure, Horizontal Inequality and Political Violence: A Disaggregated Study of Indonesian Provinces, 1990–2003." *The Journal of Development Studies* 47, no. 3 (2011): 377–398.

"The Polity Project." Vienna, VA: Center for Systemic Peace. Published 2014. http://www.systemicpeace.org/polityproject.html.

Page, Scott E. "Path Dependence." *Quarterly Journal of Political Science* 1, no. 1 (2006): 87–115.

Patel, David, Valerie Bunce, and Sharon Wolchik. "Diffusion and Demonstration." In *The Arab Uprisings Explained: New Contentious Politics in the Middle East,* edited by Marc Lynch, 57–74. New York, NY: Columbia University Press, 2014.

Petraeus, David Howell, and James F. Amos. *FM 3-24 2006 Counterinsurgency*. Boulder, CO: Paladin Press, 2009.

Pinczuk, Guillermo, ed. *Thresholds of Violence*. Fort Bragg, NC: USASOC, 2013.

Popkin, Samuel L. *The Rational Peasant: The Political Economy of Rural Society in Vietnam*. Berkeley, CA: University of California Press, 1979.

Profitt, Norma Jean. "'Battered Women' as 'Victims' and 'Survivors': Creating Space for Resistance." *Canadian Social Work Review* 13 (1996): 23–38.

Quinn, J. Michael, T. David Mason and Mehmet Gurses. "Sustaining the Peace: Determinants of Civil War Recurrence." *International Interactions* 33, no. 2 (2007): 167–193.

Raleigh, Clionadh. "Seeing the Forest for the Trees: Does Physical Geography Affect a State's Conflict Risk?" *International Interactions: Empirical and Theoretical Research in International Relations* 36, no. 4 (2010): 384–410.

Ram, Haggay. *Myth and Mobilization in Revolutionary Iran: The Use of the Friday Congregational Sermon*. Washington, DC: American University Press, 1994.

Regan, Patrick M. "Third-Party Interventions and the Duration of Intrastate Conflicts." *Journal of Conflict Resolution* 46, no. 1 (2002): 55–73.

Regan, Patrick M. and Sam R. Bell. "Changing Lanes or Stuck in the Middle: Why are Anocracies More Prone to Civil War?" *Political Research Quarterly* 63, no. 4 (2010): 747–759.

Reynal-Querol, Marta. "Ethnicity, Political Systems, and Civil Wars." *Journal of Conflict Resolution* 46, no. 1 (2002): 29–54.

Roeder, Philip G. "Ethnolinguistic Fractionalization (ELF) Indices, 1961 and 1985." *UC San Diego*. Last modified September 18, 2001. http://pages.ucsd.edu/~proeder/elf.htm.

Roessler, Philip. *Ethnic Politics and State Power in Africa: The Logic of the Coup-Civil War Trap*. New York, NY: Cambridge University Press, 2016.

Rüegger, Seraina."Conflict Actors in Motion: Refugees, Rebels and Ethnic Groups." PhD diss., ETH Zürich, 2013. https://icr.ethz.ch/publications/conflict-actors-in-motion/.

Rustad, Siri Camilla Aas, Jan Ketil Rød, Wenche Larsen, and Nils Petter Gleditsch. "Foliage and Fighting: Forest Resources and the Onset, Duration, and Location of Civil War." *Political Geography* 27, no. 7 (2008): 761–782.

Sageman, Marc. *Understanding Terror Networks.* Philadelphia, PA: University of Pennsylvania Press, 2004.

———. *Understanding Al Qaeda and the New Terror Networks.* Speech presented to the Foreign Policy Research Center, Philadelphia, PA, October 4, 2010. Accessed May 9, 2011, http://www.fpri.org/multimedia/20101004.sageman.terrorism.html.

Salehyan, Idean. "Transnational Rebels: Neighboring States as Sanctuary for Rebel Groups." *World Politics* 59, no. 2 (2007): 217–242.

Salehyan, Idean and Kristian Skrede Gleditsch. "Refugees and the Spread of Civil War." *International Organization* 60, no. 2 (2006): 335–366.

Salehyan, Idean, David Siroky, and Reed M. Wood. "External Rebel Sponsorship and Civilian Abuse: A Principal-Agent Analysis of Wartime Atrocities." *International Organization* 68, no. 3 (2014): 633–661.

Sambanis, Nicholas. "A Review of Recent Advances and Future Directions in the Quantitative Literature on Civil War." *Defence and Peace Economics* 13, no. 3 (2002): 215–243.

Savage, Sara and Jose Liht. "Mapping Fundamentalisms: The Psychology of Religion as a Sub-discipline in the Understanding of Religiously Motivated Violence." *Archive for the Psychology of Religion* 30, no. 1 (2008): 75–91.

Sawyers, Traci M. and David S. Meyers. "Missed Opportunities: Social Movement Abeyance and Public Policy." *Social Problems* 46, no. 2 (1999): 187–206.

Shapiro, Jacob N. *The Terrorist's Dilemma: Managing Violent Covert Organizations.* Princeton, NJ: Princeton University Press, 2013.

Sharp, Gene. *Sharp's Dictionary of Power and Struggle: Language of Civil Resistance in Conflicts.* New York, NY: Oxford University Press, 2011.

Silke, Andrew. "Cheshire-cat Logic: The Recurring Theme of Terrorist Abnormality in Psychological Research." *Psychology, Crime and Law* 4, no. 1 (1998): 51–69.

Skocpol, Theda. *States and Social Revolution.* Cambridge, UK: Cambridge University Press, 1979.

Snow, David A. "Identity Dilemmas, Discursive Fields, Identity Work, and Mobilization: Clarifying the Identity-Movement Nexus." In *Dynamics, Mechanisms, and Process: The Future of Social Movement Research*, editors Jacquelien von Stekelenburg, Conny Roggeband, and Bert Klandermans, 263–280. Minneapolis, MN: University of Minnesota Press, 2013.

Snow, David A. and Robert D. Benford. "Ideology, Frame Resonance, and Participant Mobilization." *International Social Movement Research* 1, no. 1 (1988): 198.

Sorel, Goerges. *Reflections on Violence*. Cambridge, UK: Cambridge University Press, 1999.

Spears, Ian. "States-within-States: An Introduction to their Empirical Attributes." In *States-within-States: Incipient Political Entities in the Post Cold War Era*, edited by Paul Kingston and Ian Spears, 15–34. New York, NY: Palgrave Macmillan, 2004.

Springer, Devin R., L. Regens, and David N. Edger. *Islamic Radicalism and Global Jihad*. Washington, DC: Georgetown University Press, 2009.

Staniland, Paul. *Networks of Rebellion: Explaining Insurgent Cohesion and Collapse*. Ithaca, NY: Cornell University Press, 2014.

Stewart, Frances. *Horizontal Inequalities and Conflict*, edited by Frances Stewart. New York, NY: Palgrave Macmillan, 2008.

Stoll, David. *Between Two Armies: In the Ixil Towns of Guatemala*. New York, NY: Columbia University Press, 1993.

Tarrow, Sidney. *Power in Movement: Social Movements and Contentious Politics*. Cambridge, UK: Cambridge University Press, 1998.

———. *Power in Movement: Social Movements and Contentious Politics*. New York: Cambridge University Press, 2011.

Taylor, Verta. "Social Movement Continuity: The Women's Movement in Abeyance." *American Sociological Review* 54, no. 5 (October, 1989): 761–775.

TC 18-01. *Special Forces Unconventional Warfare*. Washington, DC: Headquarters, Department of the Army, 2010.

Tilly, Charles. *Popular Contention in Great Britain, 1758–1864*. Cambridge, MA: Harvard University Press, 1995.

———. *Regimes and Repertoires*. Chicago, IL: University of Chicago, 2010.

Tilly, Charles and Sidney Tarrow. *Contentious Politics.* New York, NY: Oxford University Press, 2007.

Toft, Monica Duffy. *Securing the Peace: The Durable Settlement of Civil Wars.* Princeton, NJ: Princeton University Press, 2010.

———. "Territory and War." *Journal of Peace Research* 51, no. 2 (2014): 185–198.

Tollefsen, Andreas Foro and Halvard Buhaug. "Insurgency and Inaccessibility." *International Studies Review* 17 (2015): 6–25.

Trejo, Guillermo. *Popular Movements in Autocracies: Religion, Repression, and Indigenous Collective Action in Mexico.* New York, NY: Cambridge University Press, 2012.

Uppsala Conflict Data Program. "UCDP/PRIO Armed Conflict Dataset." *Uppsala University.* Last modified June 22, 2015, http://www.pcr.uu.se/research/ucdp/datasets/ucdp_prio_armed_conflict_dataset/.

Urdal, Henrik. "A Clash of Generations? Youth Bulges and Political Violence." *International Studies Quarterly* 50, no. 3 (2006): 607–629.

———. "The Devil in the Demographics: The Effect of Youth Bulges on Domestic Armed Conflict, 1950-2000." *Social Development Papers: Conflict and Reconstruction Paper* 14 (2004).

US Joint Chiefs of Staff. Joint Publication 1. *Doctrine for the Armed Forces of the United States.* Washington DC: Department of Defense, 25 March 2013.

US Joint Chiefs of Staff. JP 1-02. *Department of Defense Dictionary of Military and Associated Terms.* Washington DC: Department of Defense, 15 January 2016.

Vargas, Gonzalez. "Urban Irregular Warfare and Violence Against Civilians: Evidence From a Colombian City." *Terrorism and Political Violence* 21, no. 1 (2009): 110–132.

Varvin, Sverre. "Humiliation and the Victim Identity in Conditions of Political and Violent Conflict." *Scandinavian Psychoanalytic Review* 28, no. 1 (2005): 40–49.

Victoroff, Jeff and Arie Kruglanski. *Psychology of Terrorism: Classic and Contemporary Insights.* Washington, DC: Psychology Press, 2009.

Viterna, Jocelyn S. "Pulled, Pushed, and Persuaded: Explaining Women's Mobilization into the Salvadoran Guerrilla Army." *American Journal of Sociology* 112, No. 1 (2006): 1–45.

Vreeland, James Raymond. "The Effect of Political Regime on Civil War Unpacking Anocracy." *Journal of Conflict Resolution* 52, no. 3 (2008): 401–425.

Wallenstein, Peter and Margareta Sollenberg. "Armed Conflicts, Conflict Termination and Peace Agreements, 1989–96." *Journal of Peace Research* 34, no. 3 (1997): 339–358.

Walter, Barbara F. "The Critical Barrier to Civil War Settlement." *International Organization* 51, no. 3 (1997): 335–364.

———. "Why Bad Governance Leads to Repeat Civil War." *Journal of Conflict Resolution* 59, no. 7 (2015): 1424–1272.

Weede, Erich. "On Political Violence and Its Avoidance." *Acta Politica* 39, no. 2 (2004): 152–178.

Weidmann, Nils B., Jan Ketil Rød, and Lars-Erik Cederman. "Representing Ethnic Groups in Space: A New Dataset." *Journal of Peace Research* 47 (2010): 491–499.

Weinstein, Jeremy M. *Inside Rebellion: The Politics of Insurgent Violence.* New York, NY: Cambridge University Press, 2007.

White, Peter B., Dragana Vidovic, Belén González, Kristian Skrede Gleditsch, and David E. Cunningham. "Nonviolence as a Weapon of the Resourceful: From Claims to Tactics in Mobilization." *Mobilization: An International Quarterly* 20, no. 4 (2015): 471–491.

Wickham-Crowley, Timothy. *Guerrillas and Revolution in Latin America: A Comparative Study of Insurgents and Regimes since 1956.* Princeton, NJ: Princeton University Press, 1992.

Wimmer, Andreas, Lars-Erik Cederman, and Brian Min. "Ethnic Politics and Armed Conflict: A Configurational Analysis of a New Global Data Set." *American Sociological Review* 74, no. 2 (2009): 316–337.

Wood, Elisabeth Jean. *Insurgent Collective Action and Civil War in El Salvador.* Cambridge, UK: Cambridge University Press, 2003.

"The Worldwide Governance Indicators (WGI) Project." *The World Bank.* 2015. http://info.worldbank.org/governance/wgi/index.aspx#home.

Yu, Angela J. and Peter Dayan. "Uncertainty, Neuromodulation, and Attention." *Neuranatomy* 46, no. 4 (2005): 681–692.

Zayyat, Muntasir. *The Road to Al-Qaeda: The Story of Bin Laden's Right-Hand Man.* London, UK: Pluto Press, 2004.

ACRONYMS

ANAPO	Alianza Nacional Popular, or National Popular Alliance
ARIS	Assessing Revolutionary and Insurgent Strategies
ETA	Euskadi Ta Askatasuna, or Basque Homeland and Freedom
CSF	Central Security Forces
EAM	National Liberation Front, or Ethniko Apeleftherotiko Metopo
EIG	Egyptian Islamic Group
EIJ	Egyptian Islamic Jihad
ELAS	Greek People's Liberation Army, or Ellinikós Laïkós Apeleftherotikós Stratós
ELF	Ethno-Linguistic Fractionalization
ELN	Ejército de Liberación Nacional
FARC	Fuerzas Armadas Revolucionarias de Colombia
FLN/EZLN	Fuerzas de Liberación Nacional/Ejército Zapatista de Liberación Nacional
FMLN	Farabundo Marti National Liberation Front
GDP	gross domestic product
GIS	General Intelligence Services
IRA	Irish Republican Army
JUCO	Juventudes Comunista

LSG	Loss of Strength Gradient
LTTE	Liberation Tigers of Tamil Eelam
LURD	Liberians United for Reconciliation and Democracy
M-19	Movimiento 19 de Abril
MAK	Maktab al-Khidamat
NAACP	National Association for the Advancement of Colored People
NICRA	Northern Ireland Civil Rights Association
NRA	National Resistance Army
PIRA	Provisional Irish Republican Army
PIRA	Provisional Irish Republican Army
PKK	Partiya Karkerên Kurdistan
PLO	Palestine Liberation Organization
PRS	Political Risk Services
RENAMO	Mozambican National Resistance, or Resistencia Nacional Moçambicana
RFP	Rwandan Patriotic Front
RNA	Republic of New Africa
SORO	Special Operations Research Office
SSI	State Security Investigations

UCDP/PRIO	Uppsala Conflict Data Program/Peace Research Institute, Oslo
UN	United Nations
UNITA	National Union for the Total Independence of Angola, or União Nacional para a Independência Total de Angola
USASOC	United States Army Special Operations Command
ZAPU	Zimbabwe African People's Union

ABOUT THE AUTHORS

Summer Agan is a senior social scientist at the Johns Hopkins University Applied Physics Laboratory (JHU/APL). Her research examines organizational behavior in violent and nonviolent conflict processes. She has transitioned social science knowledge into operationally relevant education, training, concept development, and conflict modeling research with the Army Special Operation Forces community for over a decade. She holds a PhD in political science from the University of Maryland, College Park, and a bachelor's degree in international studies from the University of Washington.

Joseph (Joe) Buche is the Program Manager for the Land Power Portfolio in the National Security Analysis Department of JHU/APL. He manages a program that includes future warfare concepts and technology exploration, special operations future studies and analysis, asymmetric warfare challenges, current and future operational planning and logistics analysis, service enterprise transformation, and land domain training analysis. He also serves as a subject matter expert within JHU/APL on military capstone and operating concepts, land power, land domain training and operations, and joint military operations. Joe served for over three decades in the Army as an infantry officer. During his career he commanded four infantry companies, an infantry battalion, and an infantry brigade, in addition to serving as a deployed three star command's chief of staff. When not assigned in senior leadership or command roles, Joe worked mostly as a planner and concepts developer at the strategic, operational, and tactical levels. His assignments outside the military include serving as a division chief in the National Counterterrorism Center and as the Special Assistant to the Director of the Defense Advanced Research Projects Agency (DARPA). Joe completed professional education common to senior military officers, including selection for and graduation from the Command and General Staff College and the War College. He earned a master's degree from the elite US Army School of Advanced Military Studies (SAMS). Joe is a member of the University of Wisconsin Army ROTC Hall of Fame and a member of MENSA.

Jonathon Cosgrove is a counterforce and irregular warfare analyst at JHU/APL. He has authored and contributed to several ARIS project studies since 2014, including the Conceptual Typology of Resistance, "Little Green Men: A Primer on Modern Russian Unconventional Warfare, Ukraine 2013–2014," and others. He holds a master's degree in statecraft and national security affairs, and a bachelor's degree in political science.

Margaret (Meg) Keiley-Listermann is the Project Manager for the Assessing Revolutionary and Insurgent Strategies (ARIS) program. She additionally serves as a senior analyst at National Security Analysis Department (NSAD) where her current work includes researching training capability requirements for the U.S. Marine Corps (USMC) and conducting research and analysis for resistance education for U.S. Army Special Forces (ARSOF). A former college professor, she is published on security topics related to Northern Ireland and East Africa. Meg holds a Ph.D. in Political Science from the University of Alabama, a M.A. in Political Science from Auburn University, and a B.A. in Political Science and European Studies from Queens University of Charlotte.

Robert R. Leonhard is a research analyst at JHU/APL. He served as an infantry officer and war planner in the US Army before coming to JHU/APL. He is the author of numerous books and essays, including Fighting by Minutes: Time and the Art of War; The Principles of War for the Information Age; and the "Little Green Men" primer. He holds a PhD in American history from West Virginia University, master's degrees in international relations and military arts and sciences, and a bachelor's degree in history from Columbus University, Columbus, Georgia.

Guillermo Pinczuk is a member of the Senior Professional Staff of JHU/APL. He has authored various studies for the Assessing Revolutionary and Insurgent Strategies project, including studies focused on conflicts in Sri Lanka, Rhodesia/Zimbabwe, and the Middle East. He holds master's degrees in international affairs and applied math and statistics.

Idean Salehyan is a Professor of Political Science at the University of North Texas and the Co-Director of the Social Conflict Analysis Database project. His research interests focus on civil conflict, protest, repression, and refugee migration. He received his PhD in political science from the University of California, San Diego in 2006.

INDEX